Advanced Carpentry and Joinery for Craftsmen

Carpentry For Builders

by A. B. Emary

DRAKE PUBLISHERS INC.
NEW YORK • LONDON

Published in 1976 by
Drake Publishers Inc.
801 Second Avenue
New York, N.Y. 10017

ISBN: 0-8473-1159-7

Printed in The United States of America

Contents

Preface

This book on advanced carpentry and joinery together with its companion volume on intermediate carpentry have been designed to cover most subjects found in the syllabuses of the external examining bodies such as the City and Guilds of London Institute, Union of Lancashire and Cheshire Institutes, Union of Educational Institutes, the Incorporated British Institute of Certified Carpenters, and others.

The subject of woodcutting machinery has been omitted firstly because of insufficient space and secondly because a reference book written solely on woodcutting machinery is a much better introduction to the subject than one short chapter—all that could have been devoted to this subject. Practical geometry has its full quota of pages because it has a large part to play in the lives of woodworkers. The chapter on applied geometry includes work which is based on all the common geometrical solids. Reference has been made throughout the pages to members of the industry who have supplied information and photographs and the author is indebted to them for this.

Although many of the finer parts of the crafts of carpentry and joinery seem to be disappearing today for economic reasons and also what appears to be a lack of interest, the author feels that there will always be a large percentage of craftsmen who will still keep that pride of craftsmanship within them and who will always be able to 'turn out' a good job of work when called upon to do so.

Let us all hope that this spirit will prevail amid the coming years so that in the near future, the country and the industry will put craftsmanship first enabling us to look at modern joinery and feel that surge of pleasure and appreciation obtained only a few years ago.

I hope that this book and its companion volume will help, if only in a small way, those who seek to further knowledge in our wonderful crafts of carpentry and joinery.

A. B. Emary 1967

1 Shoring

When buildings are being repaired, or if structural alterations are being carried out, the walls of the building, and often the floors, have to be supported. Sometimes, too, walls have to be propped temporarily if they start to bulge. Raking shores have been covered in the companion volume *(Practical Carpentry and Joinery)*, but often complications arise so that the shoring has to be carried out to meet the particular conditions at the site.

Arched shoring. Fig. 1 represents the front and side elevations of a system of raking shores somewhat different from those already described. In such a case, perhaps a building on a main road must be supported, but the busy state of the roadway may induce the authorities to insist that no obstruction is placed in the way of passers-by. To avoid rakers straddling the pathway, a frame, made from 9 in. by 9 in. timbers, could form a kind of archway over the path.

The weakness in a shoring job such as this, of course, is the tendency for the top of the frame to be pushed outwards and on to the road. To prevent this a number of 2 in. stiffening boards can be bolted across the feet of the rakers and extended to the head and inside post of the frame. These stiffening boards will act as braces and tend to hold the frame in its correct shape and so overcome the outward force. The frame must be made to withstand any lateral movement parallel to the wall of the building; the side elevation shows how this can be achieved. Two 9 in. by 9 in. struts are cut to fit up against well-fixed cleats to the outside post and the sole piece. The latter should be about 10 ft. long and run along the curb to the pavement. The cleats, which should be at least 3 in. thick and 12 in. or more long, should be secured to the timbers with $\frac{3}{4}$ in. diameter coach screws. All the 9 in. by 9 in. timbers are butt-jointed to one another and the joints secured by large metal dogs.

Raker feet are cut to fit on top of the frame head and against a 3 in. cleat which also is fixed with coach screws. As the floor joists rest on the wall being supported, the centre lines of the rakers must go through to the centre of the support for the joists. The wall piece for this type of work should be 3 in. thick, and when prepared should first be secured to the wall using wall hooks (see *Practical Carpentry and Joinery)*. For additional stability it should be allowed to extend down the head of the frame and fixed to this member.

The needles pass across the tops of the rakers, through the wall piece and into the brickwork, and should be prepared from 4 in. by 6 in. material. A few stiffening boards, in addition to those at the frame, should be bolted to the rakers and wall piece to give additional stiffness. Fig. 2 shows the details around the feet of the rakers and outside corners of the frame. As an alternative, the cleats at the tops of the struts can extend up to the top of the frame.

Flying shores. This type of shoring is much more efficient than raking shores. There is always a tendency for the tops of raking shores to be pushed outwards by the forces in the wall, but in flying, or horizontal shoring as it is sometimes called, the horizontal supports are placed parallel to those forces, and so offer greater resistance. The orthodox method of erecting flying shores is seen in Fig. 7.

This type of shoring may be used, for example, when a four-storey building in a terrace has had to be demolished. While preparations are made for another building to be erected in its place, the party walls of the adjacent houses must be given support to prevent damage. The obvious method would be to erect two horizontal shores to support the walls at the floors to the third and fourth storeys, and have rakers supporting the walls at the first floor level and the ceiling level of the top floor as shown in Fig. 7. If the building had three storeys, then one horizontal shore would be fixed at the same level as the lower horizontal shore in Fig. 7, with rakers from this to go down to the first floor level and up to the ceiling level of the second floor, see Fig. 3a. Centre lines are again important (see to the right of Fig. 7).

Process of erection. To erect the flying-shore system seen in Fig. 7, holes should first be made in the walls at the correct positions to receive the ends of the needles. Next the wall pieces should be prepared and fixed by means of wall hooks, and the needles and cleats then secured.

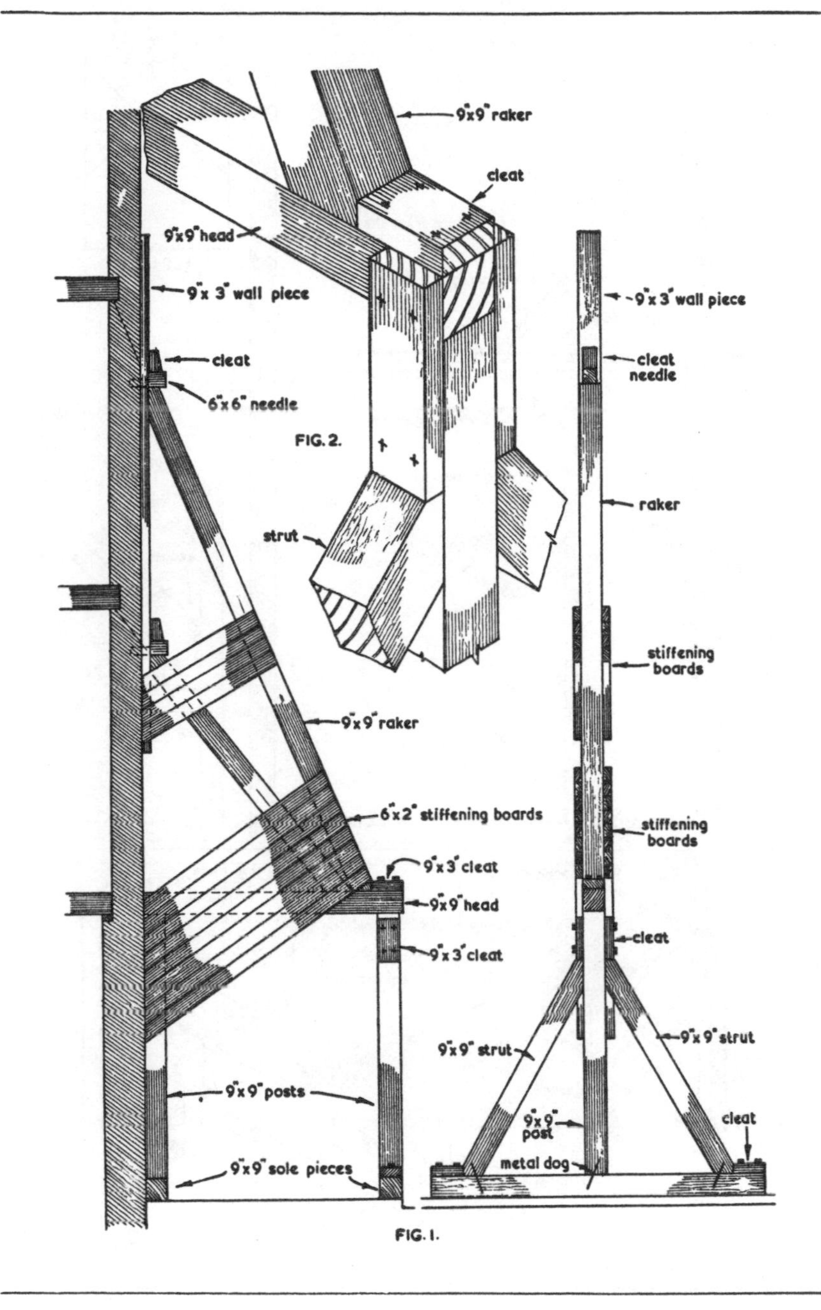
9"x9" raker
cleat
9"x9" head
9"x 3" wall piece
cleat
6"x6" needle
FIG. 2.
strut
9"x9" raker
6"x2" stiffening boards
9"x3" cleat
9"x9" head
9"x3" cleat
9"x9" posts
9"x9" sole pieces
9"x 3" wall piece
cleat
needle
raker
stiffening boards
stiffening boards
cleat
9"x 9" strut
9"x 9" strut
9"x 9" post
cleat
metal dog
FIG. 1.

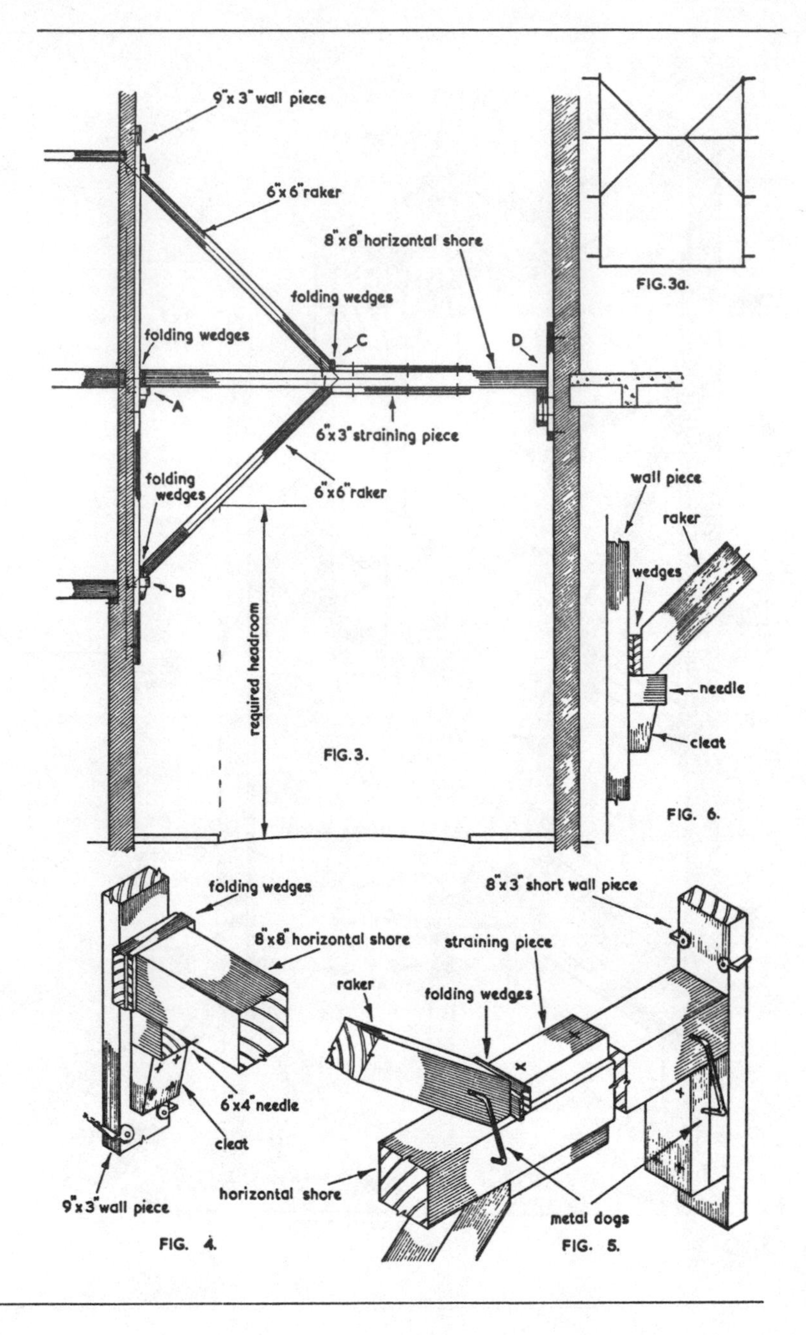

FIG. 3.

FIG. 4.

FIG. 5.

In a job of this size the cleats and needles should be prepared from 4 in. by 4 in. timber.

Next the two horizontal shores should be cut to length. These are 6 in. by 6 in. timbers offered up into position, and tightened by means of a pair of folding wedges at one end of each. The two vertical struts between the two horizontal shores should be positioned so that the rakers can be fixed top and bottom at an angle of approximately 45°. The struts can be cut from 4 in. by 3 in. material.

The four rakers are cut from 4 in. by 4 in. timbers and placed in position, the top two being tightened by a pair of folding wedges between the foot and the 4 in. by 3 in. straining piece nailed to the top of the horizontal shore. The lower two are tightened also by a pair of folding wedges placed at the foot of each as shown. A 4 in. by 3 in. straining piece is also fixed to the lower edge of the bottom horizontal shore to give the rakers an abutment. Joints should be securely fixed with large metal dogs.

Special conditions may exist for any kind of job of this sort. Fig. 3 for instance shows another system of flying shoring, where the timbers span the space across a narrow main thoroughfare, and where a certain headroom is required for passing traffic. As the wall of only one of the buildings needs support, it may be possible, if the building opposite is suitable and if permission has been obtained, to receive support from that building. It may then only be necessary to have one horizontal shore, as seen in the drawing. The building opposite has reinforced concrete floors, as shown, so it could be thought adequate to transfer all the forces from the building needing support to one point on the building opposite, in this case to a point immediately in front of a concrete floor. Only a short wall piece is needed for the building offering support. This can be fixed with wall hooks, and a double cleat to give a 4 in. to 6 in. seating for the horizontal shore is adequate at this point. Figs. 4, 5 and 6 show details at various points on the work.

Supporting an arcade. Occasionally, columns supporting arcading in Fig. 8, show signs of damage, and it may be necessary to dismantle the column completely and replace it with a new one. It is clearly necessary to support the arches immediately above the column whilst the work is being done. One method for carrying out this work is seen in Fig. 8. It consists of a yoke of two 6 in. by 6 in. timbers, each shaped on one edge to fit up against the surface of the arch with which it is in contact. The two pieces of the yoke are bolted together, as in the plan, and are supported at each end by a trestle made from 6 in. by 6 in. timbers.

The legs of the trestles rest on a sole piece with folding wedges between. The wedges are used for ensuring a tight fit for the yoke up against the arch surfaces, and if required could be placed between the yoke and top rail of the trestles instead of below the feet. To allow working space the trestles should be kept, say, 4 ft. apart. The trestle legs should preferably be splayed outwards to provide more space at floor level, but this is not absolutely necessary.

Dead shoring. This is necessary when a section of a wall of a building is removed for some purpose, such as renewing the beams over a shop front, enlarging an existing window in a building, and so on. Fig. 9 shows an elevation of a building which is having a large window incorporated in the ground floor external wall. A glance at the section through the building, Fig. 10, shows that the first and second floor joists are carried by the wall in which the alterations are taking place. It is first necessary to relieve the wall of these loads, and this is done by propping up the joists from a solid base with 6 in. by 4 in. timbers placed approximately 5 ft. apart across the front of the building. These props, of course, must have 6 in. by 3 in. heads and sole pieces, with folding wedges under each to tighten the work up efficiently.

The next decision is whether to give support to the outside surfaces of the external walls around where the work is to be carried out. If it is decided that this support is necessary, it can be in the form of raking shores as seen in the two drawings. It may also be considered that support should be given to the openings in the wall immediately above the alterations. In this case these are window openings, and the sides are supported to prevent damage to the brickwork. The sashes should be removed and the frames given extra strength by introducing side pieces and props to each opening. It may also be necessary to introduce props in an adjoining door opening as can be seen at ground floor level in Fig. 9.

The next step is to decide where the dead shoring should be positioned. They should be placed in the centres of the brickwork between the window openings, because the weight of the wall above the first floor openings is being transferred through these points.

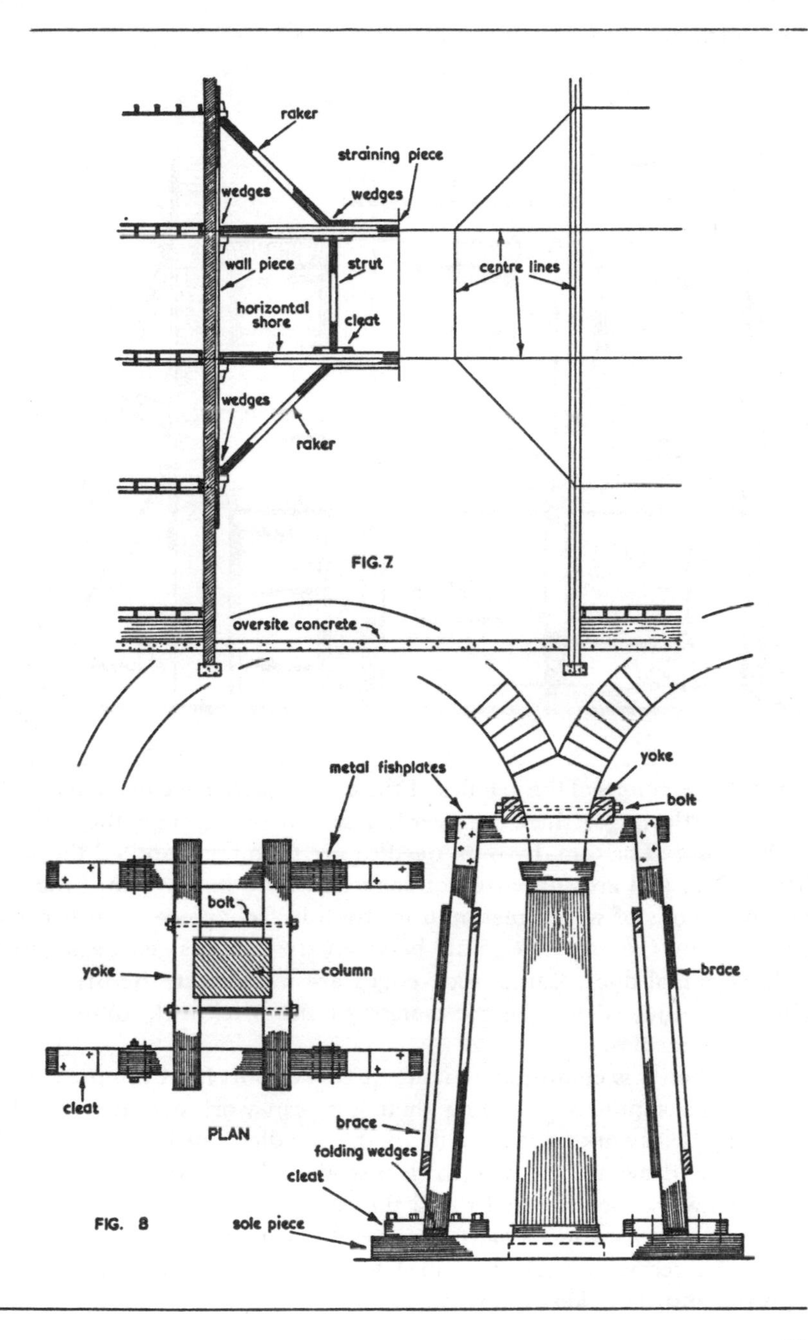
raker
straining piece
wedges
wedges
wall piece
strut
horizontal shore
cleat
centre lines
wedges
raker
FIG. 7
oversite concrete
metal fishplates
yoke
bolt
bolt
yoke
column
brace
cleat
PLAN
brace
folding wedges
cleat
FIG. 8
sole piece

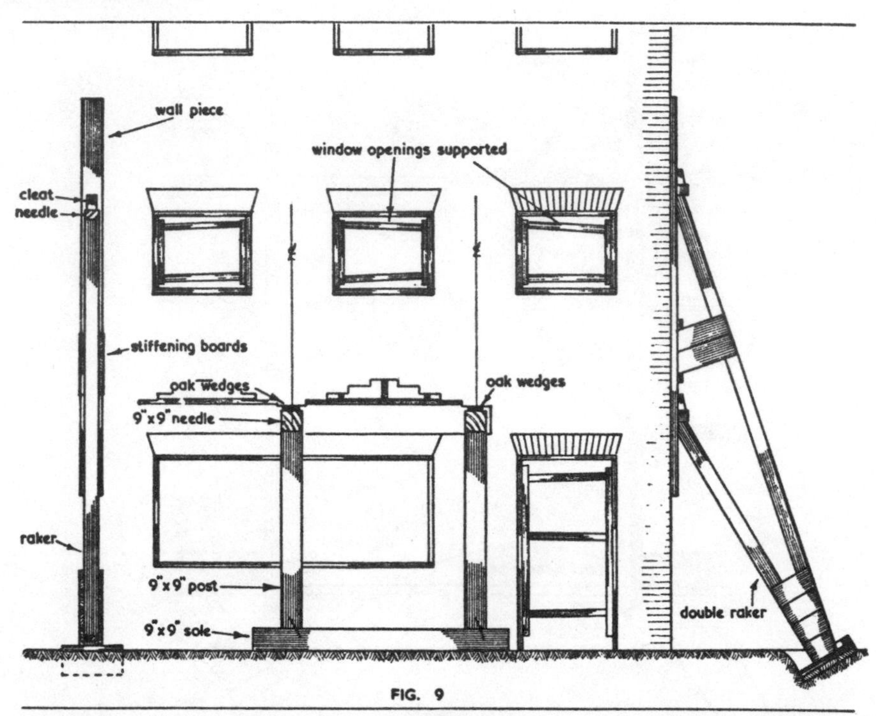

FIG. 9

When the positions of the needles of the dead shores have been decided, holes, a little larger than the needles, are made through the wall of the building. The 9 in. by 9 in. needles are then manhandled through these holes, and are supported at each end by 9 in. by 9 in. timbers, the lower ends of which rest on 9 in. by 9 in. sole pieces well bedded down on the ground. The joints between the timbers can be secured by large metal dogs. Large oak wedges are inserted in the brickwork holes and adjusted to take the weight of the brickwork immediately above the needles.

Only when it is confirmed that adequate support has been provided, and that all supports are doing their job, can work commence. The necessary brickwork on each side of the needles can be removed, as seen in the drawing, the new work introduced into the opening, and new brickwork incorporated to fill the space made by the removal of the original brickwork. Lastly the needles can be removed and the holes left filled with bricks to complete the work. Afterwards the floor supports and the rakers are removed.

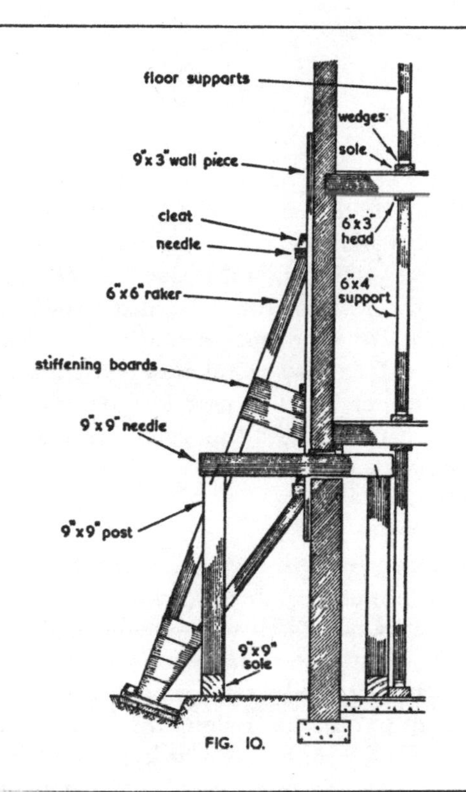
floor supports
wedges
sole
9"x 3" wall piece
cleat
needle
6"x3" head
6"x6" raker
6"x4" support
stiffening boards
9"x 9" needle
9"x9" post
9"x9" sole
FIG. 10.

2 Timbering to excavations

Many accidents, some serious, have been caused by thoughtlessness on someone's part when the sides of trenches and the like have collapsed and trapped workmen. When one realises that a cubic yard of soil can weigh up to a ton or more, it doesn't need much imagination to realise what the results would be if a person were walking down a trench 8 ft. to 10 ft. deep when the sides caved in. Risks should never be taken. It is easy to take the simplest way out when the ground around a trench looks firm and solid. 'We don't need any supports to that one' can be said all in good faith, but with serious, perhaps fatal, consequences.

When supports are needed. An excavation in the ground more than 3 ft. 6 in. deep should be supported. Fig. 1 shows the type of supports needed in firm hard ground which will remain in that state. Heavy lorries passing near by, continuous bad weather, and other conditions can cause havoc to what looks like safe ground. Poling boards, 1½ in. thick, can be placed up to 6 ft. apart along the length of a trench, and these can be supported by 3 in. by 3 in. or larger struts. If the trench is fairly shallow, say up to 3 ft. 6 in. deep, only one strut to each pair of poling boards need be used.

Where firm hard gravel is being excavated, but there is a danger of the ground becoming less stable by reason of nearby traffic, etc., extra poling boards can be introduced as in Fig. 3. So as to keep good working space in the trench, waling pieces of 4 in. by 3 in. section can be introduced, and the struts kept at a distance of 6 ft. apart. Shallow trenches, again, can be supported by single struts, and to ensure that the struts keep their positions, the sides of the trench can be tapered slightly, as seen in Fig. 4.

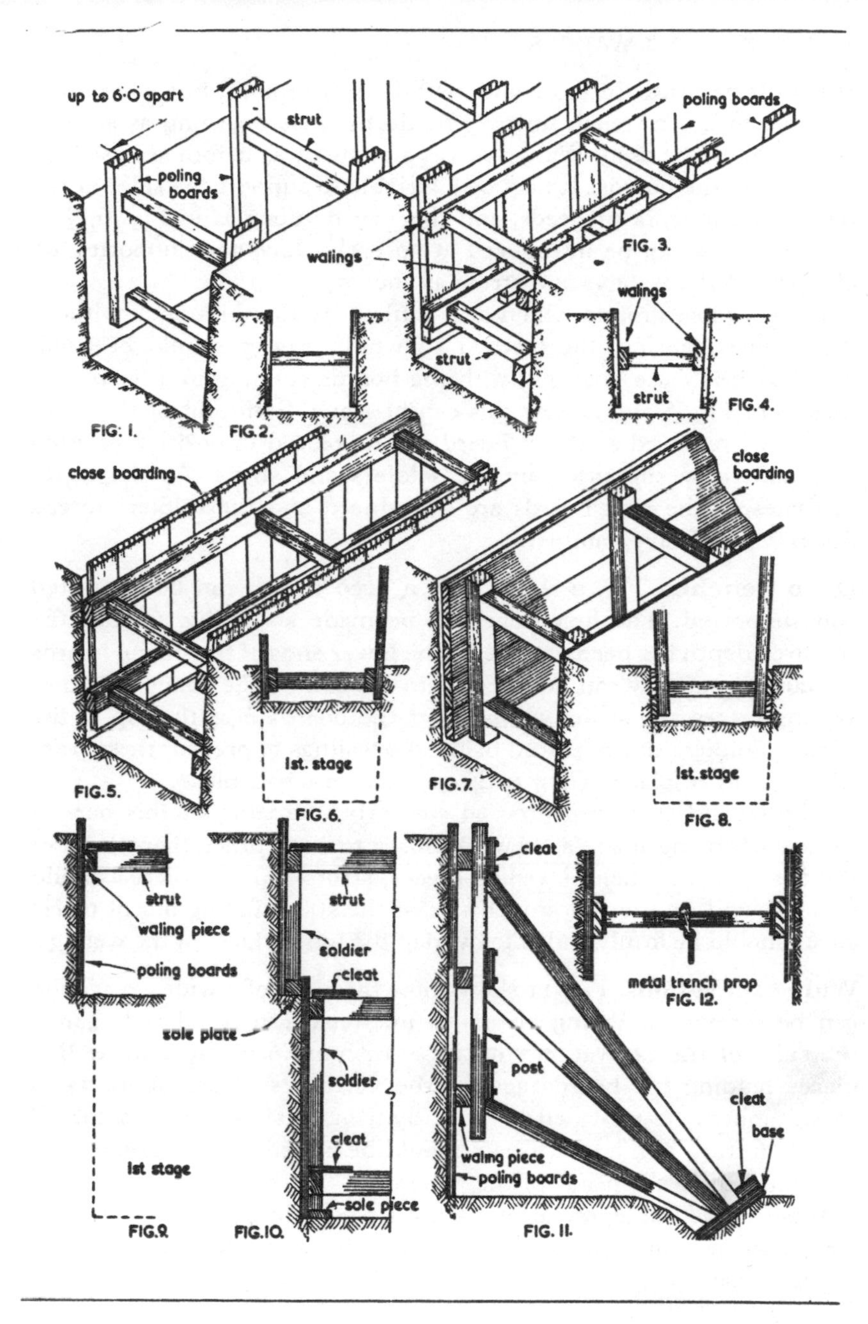
up to 6·0 apart
strut
poling boards
poling boards
walings
FIG. 3.
walings
strut
strut
FIG. 4.
FIG. 1.
FIG. 2.
close boarding
close boarding
1st. stage
FIG. 5.
FIG. 6.
FIG. 7.
1st. stage
FIG. 8.
strut
waling piece
poling boards
1st stage
FIG. 9.
strut
soldier
cleat
sole plate
soldiers
cleat
sole piece
FIG. 10.
cleat
post
waling piece
poling boards
cleat
base
FIG. 11.
metal trench prop
FIG. 12.

Difficult ground. In loose and waterlogged ground, but where it is possible to dig a trench up to 3 ft. deep, close boarding as seen in Fig. 5 should be used. Waling pieces and struts as before are used for supporting the boards. It may be advisable to support the sides of the trench as the work proceeds, and this can be done as in Fig. 6. The poling boards can be introduced at an early stage and supported by one part of the walings and struts as shown.

In very loose ground, where it is difficult to dig the trench without the sides collapsing, the method shown in Fig. 7 should be used. This again is close boarding with the boards running in a horizontal direction. Fig. 8 shows how this can be done. Sufficient of the earth should be removed so that a board can be placed in position on each side and these supported by the soldiers and struts. As the depth increases so the next boards are introduced and the soldiers forced down with a heavy hammer.

Deep trenches. Fig. 9 shows how a deep trench can be excavated and supported. The first stage can be made as in Fig. 6 until the required depth has been obtained. The lower ends of the poling boards are cut so that they can be forced into the soil by the heavy hammer. Waling pieces and struts will support the boards near the top of the trench. Soldiers can be placed below the walings to prevent these from slipping out of place. Their ends can rest on a sole piece.

The second stage can proceed similarly, the sides of this part of the trench being made somewhat closer to each other than those of the first stage. As before, waling pieces, struts and sole pieces should be used, and, to prevent any chance of the struts falling out of place, cleats should be firmly nailed to overlap the top surfaces of the walings.

Wide excavations. Fig. 11 shows how the sides of a wide excavation can be supported. Poling boards, 2 in. thick, can be placed against the sides of the excavation with three or more 6 in. by 4 in. waling pieces holding the boards against the vertical surface. Supports to these timbers are provided by 4 in. by 4 in. posts and rakers spaced at, say, 6 ft. to 8 ft. apart. Cleats should be bolted to the posts to give an abutment to the heads of the rakers.

Where practicable, metal props can be used in place of the timber struts, as seen in Fig. 12. These are easily adjusted and are more efficient than the struts.

3 Centres for Arches

Although arch work is fast disappearing in the building industry, it is still necessary for carpenters to be able to do this type of work because from time to time jobs of this nature do come along. The methods used for constructing the small arch centre were explained in the companion volume, *Practical Carpentry and Joinery*. Here we explain how large arch centres can be manufactured, and also the geometry and setting out of more difficult examples.

Semicircular arch. Fig. 1 is the elevation of a semicircular arch of 10 ft. span in a wall 2½ bricks thick. On the left is shown how to overcome a projection at the springing line. The centre consists of two 7 in. by 2 in. ties, ribs from 7 in. by 2 in. material, 6 in. by 2 in. struts, and ¾ in. thick plywood gusset plates. The lagging which is nailed to the outside edges of the ribs is 2 in. by 1½ in. All timbers are butt-jointed and the joints are secured with ¾ in. thick gusset plates with ½ in. bolts and 2 in. square washers each end.

The centre is supported on three pairs of legs, each pair being secured together as a unit by means of 6 in. by 1½ in. battens. Folding wedges are placed under the supports for final adjustment and for ease of dismantling. The supports should also be braced.

Fig. 2 is a vertical section through the centre and shows that the ribs are cross-braced. Note also that the length of the lagging pieces is about 1 in. less than the thickness of the wall, and the ribs are kept about ½ in. from the ends of the lagging.

Semi-elliptical arch. Fig. 3 is the elevation of a semi-elliptical arch of 15 ft. span, and from front to back of any dimension. This type of archway could occur at the entrance of a building leading into a central quadrangle. Again all the timbers have been connected at the

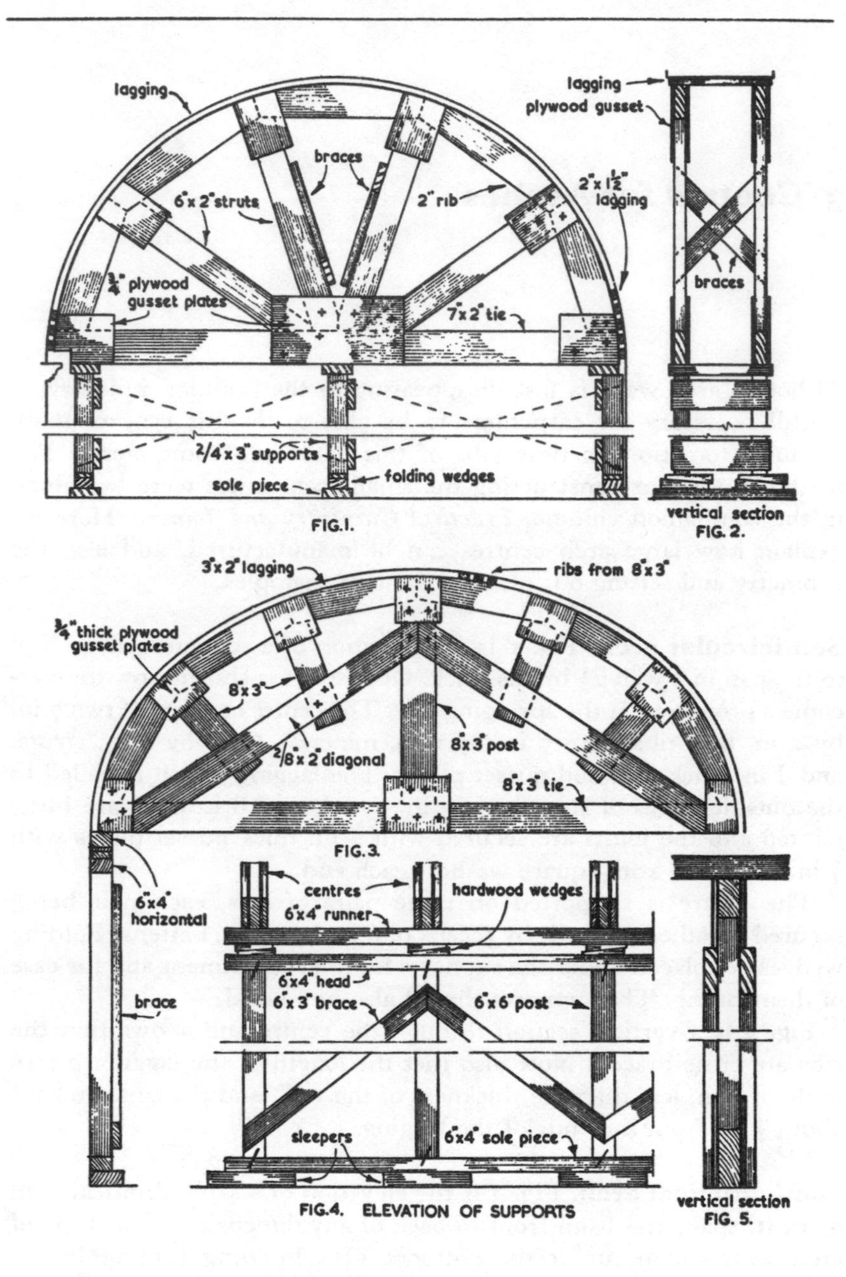

lagging
braces
6"x 2" struts
2" rib
2"x 1½" lagging
¾" plywood gusset plates
7"x 2" tie
2/4"x 3" supports
folding wedges
sole piece
FIG. 1.
lagging
plywood gusset
braces
vertical section
FIG. 2.
3"x 2" lagging
ribs from 8"x 3"
¾" thick plywood gusset plates
8"x 3"
8"x 3" post
2/8"x 2" diagonal
8"x 3" tie
FIG. 3.
6"x4" horizontal
brace
centres
hardwood wedges
6"x4" runner
6"x4" head
6"x 3" brace
6"x 6" post
sleepers
6"x4" sole piece
FIG. 4. ELEVATION OF SUPPORTS
vertical section
FIG. 5.

intersections with ¾ in. thick gusset plates and bolts. Two pairs of diagonals support the joints round the ribs and the lagging is 3 in. by 2 in. The supports consist of a frame at each side of the opening, running from front to back. These frames are made from fairly heavy timbers and the joints are secured with metal dogs. The frames should be braced, see Fig. 4.

These supporting frames should be levelled, and can be placed on timbers such as old railway sleepers. A 6 in. by 4 in. runner is placed on the head of each frame, with hardwood folding wedges between these timbers and immediately over the posts of the frames. The centres are placed on the runners and over the frame posts which can be up to 5 ft. apart. The centres should be well braced through their centre posts. The wedges are used for final adjustment and dismantling.

Arch with splayed jambs. Fig. 6 shows the plan and elevation of an elliptical arch centre for an opening which has splayed jambs and a level crown. Given the shape of the ribs on the narrow side of the opening, the problem is to develop the shape of the ribs on its wide side. Draw the plan of the opening and place in the drawing the positions of the front and rear ribs. Also draw the elevation of the outline of the given rib. More details have been placed in this portion of the drawing than is actually necessary for geometrical purposes to show how the ribs will appear when made.

Divide the plan of the given rib into any number of equal parts, say eight, and project these points up to the curve in the elevation to give points 1–8. Project the edge of the jamb downwards to meet the centre line in X, and using this as a focal point, draw lines through points 1 to 8 in the plan to give points 1′–8′ on the edge of the required or rear rib. From these points draw vertical lines, and make these the same length as those on the elevation of the rib in elevation; for instance 8′–8″ should be the same length as 8–8 line, 7′–7″ the same as 7–7 line, and so on. A freehand curve through all these points will give an outline of the shape of the other rib.

Arch in curved wall. Fig. 8 is the plan and part elevation of a centre for a semicircular-headed opening in a curved wall. In other words it is a centre based on double curvature work. This drawing and the double curvature drawings in Chapter 19 should be compared. A

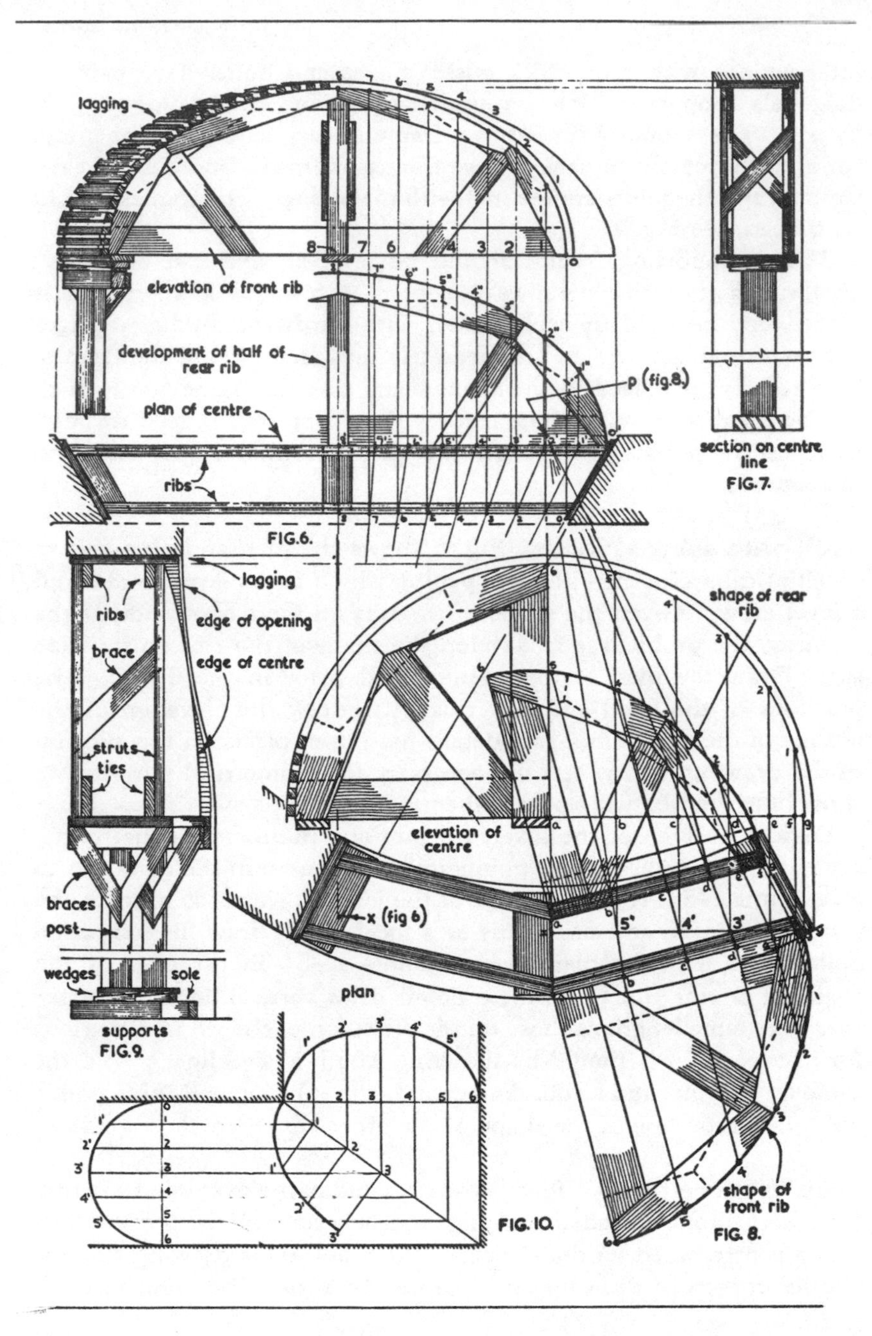
lagging
elevation of front rib
development of half of rear rib
plan of centre
ribs
FIG.6.
section on centre line
FIG.7.
lagging
ribs
edge of opening
brace
edge of centre
struts
ties
braces
post
wedges
sole
supports
FIG.9.
shape of rear rib
elevation of centre
x (fig 6)
p (fig.8.)
plan
shape of front rib
FIG. 10.
FIG. 8.

glance at the plan shows that the wall is curved and that the jambs of the opening radiate back to point P.

The semicircular curve, which has been divided up into six equal parts, is the shape of the opening on line x–y in the plan. The positions of the ribs (the front and rear ribs are each made in two halves) are seen in the plan. To construct the shapes of the front and rear ribs, first divide one half of the elevation curve into, say, six equal parts, and project these points down vertically to the x–y line in plan, to give points 5′, 4′, 3′, etc. From these points draw lines to give points a, b, c, d, etc. on the front and rear ribs, using point p as a focal point.

Draw lines from points a, b, c, etc. on the front and rear ribs, perpendicular to the plan edges of each, and make the various perpendicular lines equal in length to those in the elevation. For instance, a–6 on the front and rear ribs should be equal in length to a–6 in the elevation. A curve through each set of points, 1, 2, 3, etc. will give an outline of the ribs required for the centre. A view of the ribs developed is shown in the drawings. Fig. 9 gives a vertical section through the centre.

Fig. 10 shows the geometry involved in setting out the shapes of the centres for the ceilings to passage-ways in the basement of a building. Given the semicircular shape of one, the other, which is wider, can be developed in the way shown.

4 Gantries

Gantries are used where building operations are taking place, and are erected over the pathway. They provide a loading and unloading platform for materials, and provide a base for scaffolding without obstructing the pavement immediately in front of the building. The platform enables the builder to store old brick rubble and other materials awaiting collection without causing inconvenience to passers-by.

Main frames. A gantry, Figs. 1 and 2, consists of two frames made from large timbers, such as 7 in. by 7 in. up to 12 in. by 12 in., butt-jointed together and secured with large metal dogs. One frame rests up against, or is very near to the building, and the other rests on the pavement by the curb. To keep the frames square, braces of, say, 6 in. by 3 in. material should be fixed by means of bolts or coach screws as shown in the elevation. The two frames should also be braced together across the width of the pathway as seen in Fig. 2.

Platform. The platform consists of 3 in. thick joists laid across the heads of the frames, and 2 in. or 3 in. thick boards to form the surface of the platform. As workmen will be carrying out certain work on the platform a safety rail should be provided round the three sides about 3 ft. 6 in. high to prevent possible accidents. If there is any risk of brick rubble or any other material falling down on to the pavement below, and this risk is always present if the platform is to be used for storing materials, close boarding should be fixed to the posts holding the safely rail. This boarding needs to be about 18 in. high. It may be an advantage to have one section of the safety rail with its close boarding easily removable so that the loading and unloading of lorries can be carried out readily.

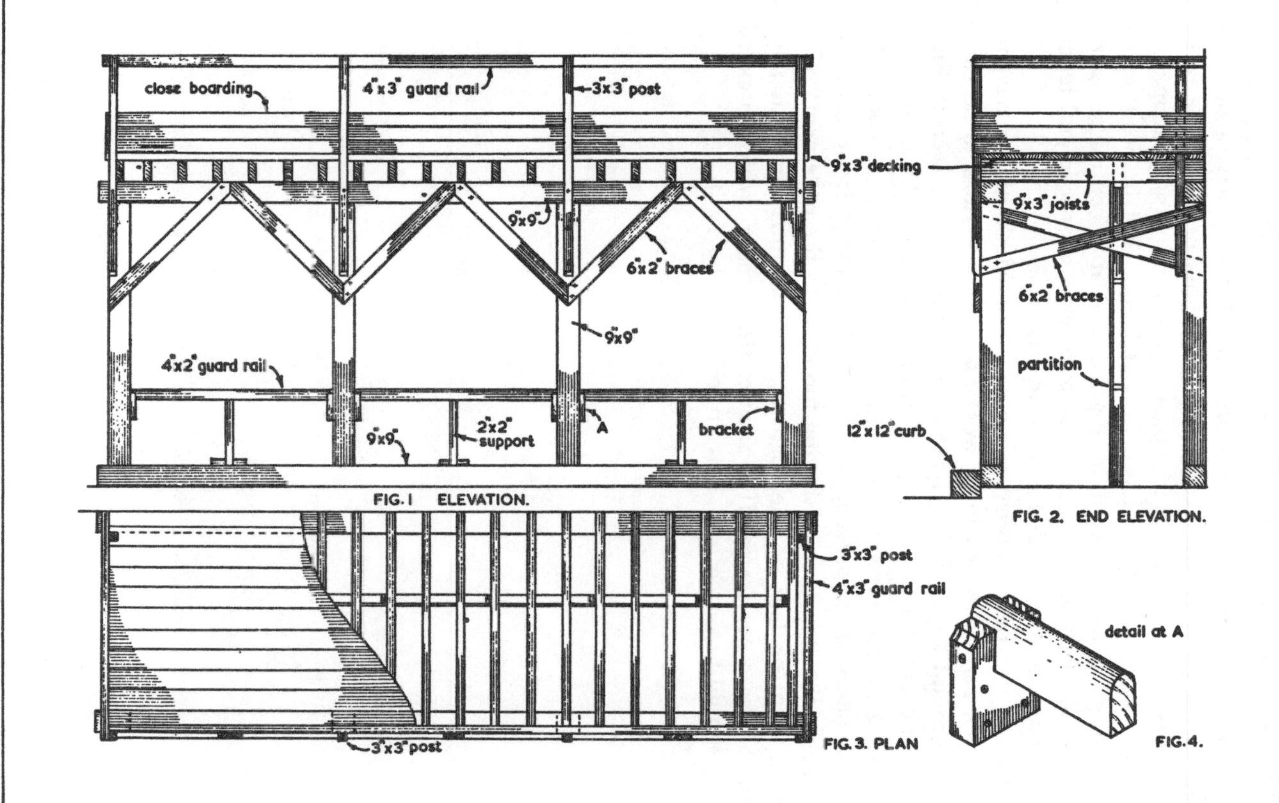

FIG. 1 ELEVATION.

FIG. 2. END ELEVATION.

FIG. 3. PLAN

FIG. 4.

As the public will be passing underneath the gantry, a safety rail should also be provided between the uprights of the frame nearest the roadway. This will stop foot passengers unthinkingly stepping out into the roadway. The rail can be fixed and held in position with brackets, such as that seen in Fig. 4, and if a central support is required for each section, a 2 in. by 2 in. post, morticed and tenoned to the rail, can be provided as seen in Fig. 1.

It may also be necessary to narrow the width of the pavement below the gantry to provide a working space for work to be carried out at ground level. If this is required, a partition, as seen in Fig. 2, should be erected at the predetermined position. This consists of 3 in. by 2 in. timbers faced with $\frac{3}{8}$ in. external-grade plywood. Often it is also necessary to prevent vehicles on the roadway from accidentally damaging the structure by driving too close to the pavement. A large timber may be laid on the roadway, in the gutter immediately in front of the outer frame. If available, this timber, which is called a curb, should be either 9 in. by 9 in. or 12 in. by 12 in.

5 Formwork for Concrete

Reinforced concrete is being used more than ever at the present time, and in the foreseeable future its use will continue to increase. The carpenter who is not specially skilled in formwork, is sometimes called upon to design and erect the formwork for an odd job. It can be difficult for this man to imagine what the required 'box' or formwork will look like in relation to the finished piece of work. Fairly simple concrete formwork has already been covered in the companion volume, *Practical Carpentry and Joinery*, and it is the object of this chapter to go a little farther into the subject to help those not particularly skilled in this field.

Essential considerations. There are certain points to keep in mind when designing formwork for concrete.
1. The formwork is really a casting box into which a semi-fluid material is poured, and is shaped in accordance with the details obtained from the working drawings.
2. The boards or sheet material forming the box must be strong enough to avoid distortion or bending under the weight of the concrete.
3. The supports to the formwork must also be strong enough to support the formwork, the concrete, and possibly the weight of workmen and barrows.
4. The weight of the reinforcement must also be taken into consideration.
5. The striking of the formwork and its supports must be considered, too, so that the timber involved will not be unnecessarily damaged.

Materials. The timber used for shuttering is often some cheap form of softwood such as white deal. Resin-bonded plywood is being used widely for the formwork surfaces to floors, beams, columns, etc.

Metal formwork, too, is being used more extensively than it was twenty years ago. Although new systems are being introduced and the writer realizes how important these are to the industry, he thinks it is still necessary for the carpenter to be able to carry out this work in timber. If he can erect formwork in timber, he certainly should be able to carry out work with metal formwork systems if requested to do so.

Where a good smooth finish to the concrete is required, ½ in. thick resin-bonded plywood sheets, adequately supported, should be used in place of boards. Alternatively, $\frac{3}{16}$ in. thick tempered hardboard can be used to line or surface the boards. Where boards only are to be used to achieve a fairly good surface the boards should be prepared on a planing machine. When a good surface is not necessary, rough boarding is used.

If possible, all formwork should be manufactured in the joiner's shop where machines are available, and if the formwork is to be used more than once this should be kept in mind so that the removal of the timber can be carried out with the minimum of damage. Often mineral oils are used to prevent the concrete from adhering to the timber. No formwork should be removed without the proper authority. Many accidents have occurred as a result of removing supports too soon.

Staircase formwork. Fig. 1 shows the plan of a concrete staircase of two flights connected by a half-space landing. Obviously, the first stage in this work is to provide the formwork and supports to the lower flight and the landing, Fig. 2.

Landing timbers should be erected first. These comprise 1¼ in. thick boards supported every 16 in. along their lengths by 3 in. by 2 in. joists. In turn the joists are supported by 6 in. by 3 in. runners and 4 in. by 3 in. posts. The posts rest on a sole piece with folding wedges between. These wedges will allow the timbers to be adjusted to the correct height and also allow the easy dismantling of the formwork when the time arrives.

The soffit of the first flight is of 1¼ in. boards, and these are supported every 16 in. on 3 in. by 2 in. joists. The last named are supported by three 4 in. by 2 in. runners, one near each side of the stairs and one in the centre. Posts, 4 in. by 3 in., positioned at right angles to the runners, transfer the weight of the concrete down to the sole piece on

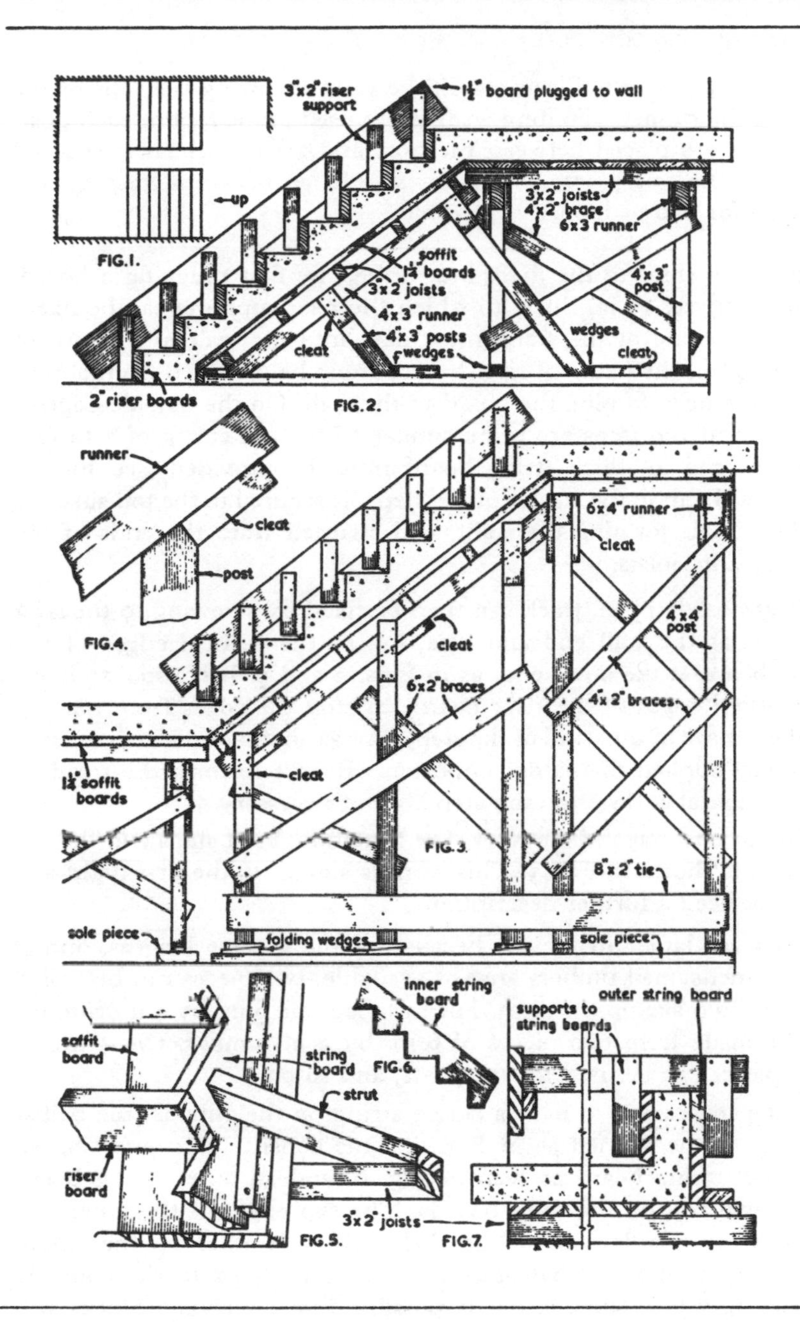
3"x2" riser support
1½" board plugged to wall
up
FIG.1.
soffit
1¼" boards
3"x2" joists
4"x3" runner
4"x3" posts
cleat
wedges
2" riser boards
FIG. 2.
3"x2" joists
4"x2" brace
6"x3 runner
4"x3" post
wedges
cleat
runner
cleat
post
FIG.4.
6"x4" runner
cleat
4"x4" post
cleat
6"x2" braces
4"x2" braces
1¼" soffit boards
cleat
FIG. 3.
8"x2" tie
sole piece
folding wedges
sole piece
inner string board
supports to string boards
outer string board
soffit board
string board
FIG.6.
strut
riser board
3"x2" joists
FIG.5.
FIG.7.

which the posts rest. Cleats should be used for fixing the tops of the posts to the runners. Folding wedges are used at the foot of each post, and these are placed between the post and a cleat which is secured firmly to the sole piece. The latter should go back and rest against the wall for extra support, if possible.

Steps. The steps to the flight are formed by first plugging a board, to the wall, say 1½ in. thick, to which the riser supports can be fixed. As the concrete of the stairs must pass into the wall to a depth of 4 in. to 6 in., the fixing of riser boards must be provided from above. Hence the need to plug the board to the wall. On the outside edge of the stairs, if the steps are to be similar to the cut string of a timber staircase, a 2 in. thick string board must be provided, see Fig. 5. This board, cut to the shapes of the steps, is secured to the top surfaces of the boards forming the soffit, and strutted from the ends of the 3 in. by 2 in. joists.

Riser boards 1½ in. thick can then be fixed by screwing to the riser supports at the wall end and screwing to the vertical edges of the string board at the outer end, as in Figs. 2 and 5. It is usual to bevel the bottom edges of the riser boards to stop the edges from making a hollow mark in the tread of the steps, but as another finish to concrete is usually applied this is not important. Boards to form the tread or the horizontal surface to each step are not necessary.

The second stage of the work, the second flight of steps and the top landing, is shown in Fig. 3. This work is similar to the first flight and does not need a further description.

Posts are larger in section because of the greater lengths, and if these dimensioned timbers are not available, two pieces can be nailed together to make up the sizes. For instance, the 6 in. by 4 in. runners can be made from two pieces of 6 in. by 2 in. timber, two 4 in. by 2 in. pieces can be used for the posts, and so on.

If the staircase is to have a closed string on the outside, the balustrading can be formed as in Fig. 7. This consists of supporting an inner-cut string board as seen in Fig. 6, and an outer string board which will rest on the soffit boards. The top edges of the inner and outer string boards will run parallel with each other up the length of the flight. If metal baluster rails are to be fixed to the concrete balustrading at a later date it is possible that dovetailed blocks will

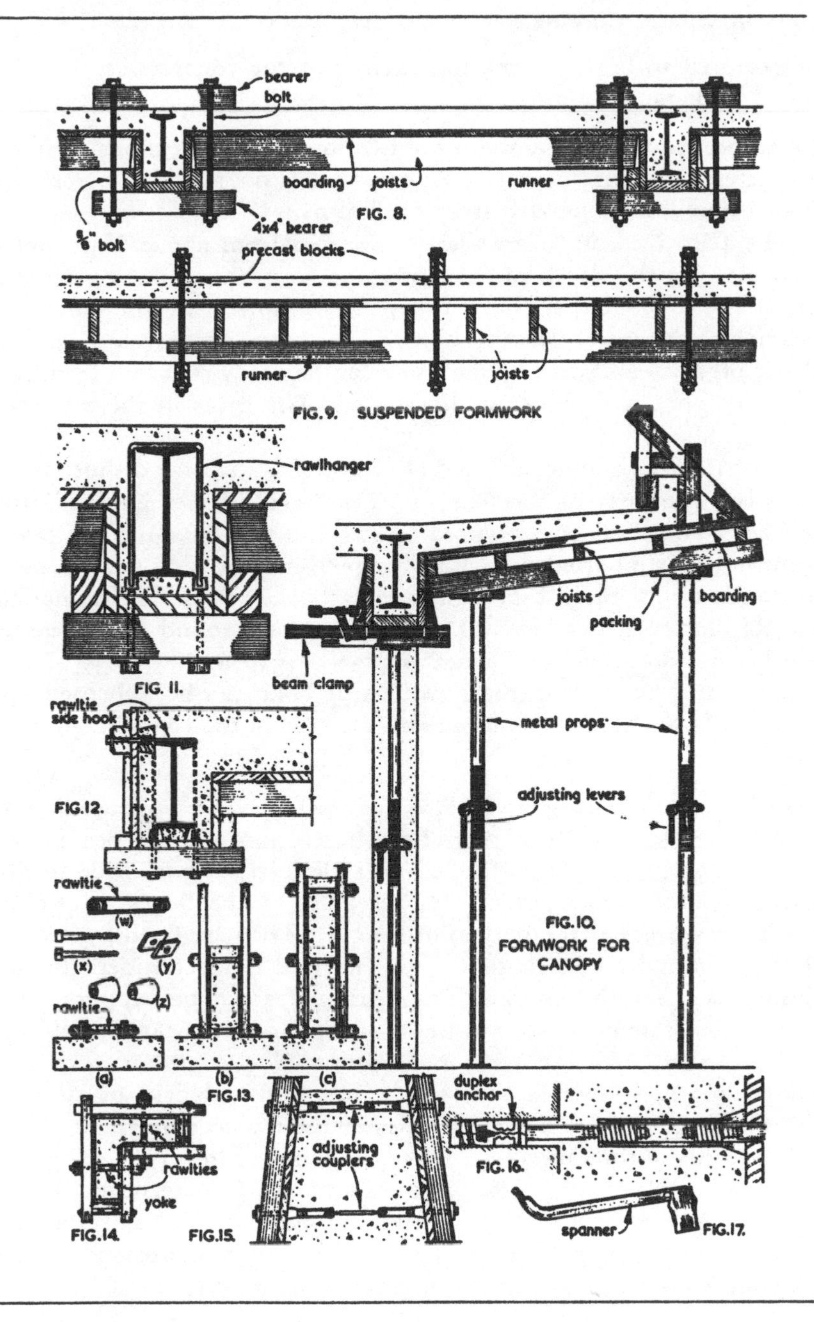
bearer
bolt
boarding
joists
runner
4x4" bearer
5/8" bolt
FIG. 8.
precast blocks
runner
joists
FIG. 9. SUSPENDED FORMWORK
rawlhanger
FIG. 11.
beam clamp
joists
packing
boarding
metal props
adjusting levers
rawltie side hook
FIG. 12.
rawltie
(w)
(x)
(y)
(z)
rawltie
(a)
(b)
(c)
FIG. 13.
FIG. 10. FORMWORK FOR CANOPY
duplex anchor
rawlties
yoke
adjusting couplers
FIG. 16.
FIG. 14.
FIG. 15.
spanner
FIG. 17.

be required to be let in the top surface of the concrete so that the metal work can be grouted in.

Concrete floor. Figs. 8 and 9 are two longitudinal sections through a concrete floor with beams. The formwork is not supported from the floor below but suspended from the beams.

The 4 in. by 4 in. bearers are supported from above. Two bolts pass through the depth of the timbering to a bearer resting on the metal beam. The upper bearer is kept to the correct height above the beam by a precast concrete block. The bolts are covered with a sleeve of, say, cardboard so that they can be easily removed from the concrete when the timber is dismantled. The holes in the concrete are then filled.

Fig. 11 shows another method of supporting suspended shuttering. This is by the use of Rawlhangers. Fig. 12 illustrates how to form the side wall to a floor supported by suspended formwork. The main formwork is supported with Rawlhangers and the side shutter held in position with the use of Rawltie side hooks. These are similar to the Rawlhangers, but have a hook which passes round the flange of the beam. The dovetailed block is placed against the flange of the beam so that the bolt in the Rawltie side hook can be tightened up. Afterwards the block is removed and the hole in the concrete grouted in.

Concrete canopy. Fig. 10 is a view of the formwork and supports to a reinforced concrete canopy. The concrete forming the beam over the opening in the wall is formed by a three-sided box made to the required dimensions. Notice that the sides of the box go down to the lower surface of the bottom of the box. They are fixed by nailing through into the box bottom, thus strengthening considerably the timber used for the bottom. The beam sides can be supported by metal beam clamps as seen in the drawing, and the beam is held up at the correct height by using metal props. No wedges are necessary under the props because they can be adjusted in height by turning the adjusting levers which raise or lower the top section.

The soffit of the canopy is formed by using 1¼ in. to 1½ in. boards supported by 3 in. by 2 in. joists spaced at 16 in. centres. These joists are carried on 4 in. by 3 in. runners at about 4 ft. centres which in turn, are supported on metal props. The outside upstand to the concrete canopy can be formed in the manner shown.

Concrete wall. Fig. 13a, b and c show a method of constructing a concrete wall with climbing formwork. Rawlties are used for these operations, and Fig. 13a shows the first step. To form the foundation to the wall a small upstand is fixed with Rawlties. The second stage, b, shows the wall taken up to the first lift, again using Rawlties and incorporating the ties that were used for forming the upstand. The next stage is seen in 13c. Formwork for the second stage has been raised so that more Rawlties are needed at the top of this next lift. Packing pieces are used to keep the bearers at the correct distance at the bottom.

A fourth stage, not shown, would entail the formwork being raised for the next layer of concrete to be poured, the top two Rawlties of the third stage being used as well as an additional one at the top of the new 'lift'. Fig. 13w, x, y and z show the various parts of a Rawltie. They consist of the welded section with the coiled steel sockets which form the thread for the bolts to be turned into; the two bolts, x; two large square washers, y; and z, the two hardwood cones. Only the welded portions are not recovered after use. Bolts, cones and washers can be used again with additional welded portions. Figs. 14 and 15 show how Rawlties can be used for formwork which would prove unsatisfactory if the welded parts were not available. Fig. 14 shows their use in a column of uncommon shape, and Fig. 15 illustrates how they can be used in conjunction with couplers. These last named can be used for adjusting the lengths of the fixture for a wall with tapered sides.

Fig. 16 illustrates how a Rawltie can be used with a Duplex anchor to fix formwork adjacent to an existing masonry wall. Fig. 17 shows the type of box spanner made for use with Rawlties, etc.

6 Roofs

Many advances have come about since the second world war in respect of roofing, due no doubt, to the cost of timber and the necessity of producing roofs of large span without intermediate supports with small section timber. These advances have covered the whole field of roofing from domestic buildings to factories. Roofs as far as domestic buildings are concerned tend to have flatter pitches than before, and many new trusses have been designed as a consequence, but, even so, the traditional pitched roof of 40° to 45° has not been forgotten in the attempt to conserve timber.

Perhaps the chief influence in these developments is the *Timber Research and Development Association* (TRADA) known originally as *The Timber Development Association* (TDA). This association has carried out research in many fields in which timber is the main material, and roofing has been one of their main subjects. The TDA truss (described in the companion volume, *Practical Carpentry and Joinery*), revolutionised roofing for domestic buildings, this truss making a far superior job compared with the old traditional method.

Many of the roof trusses described in the first part of this chapter have been designed by TRADA, and further information and drawings on many of these trusses and some others can be obtained from TRADA, Hughenden Valley, High Wycombe, Bucks.

Low-pitched roof (1). Fig. 1 shows the elevation of a low-pitched domestic truss, comprising principal rafters made from two pieces of 4 in. by 1½ in., and a lower chord made also from two pieces of 4 in. by 1½ in. Fig. 2 shows the joints at the intersection between the principal rafters and the lower chord. Three gusset plates are used, each 18 in. by 4 in. by 2 in., with three ½ in. diameter bolts and four 2 in. double-toothed timber connectors to each bolt, see Fig. 2a.

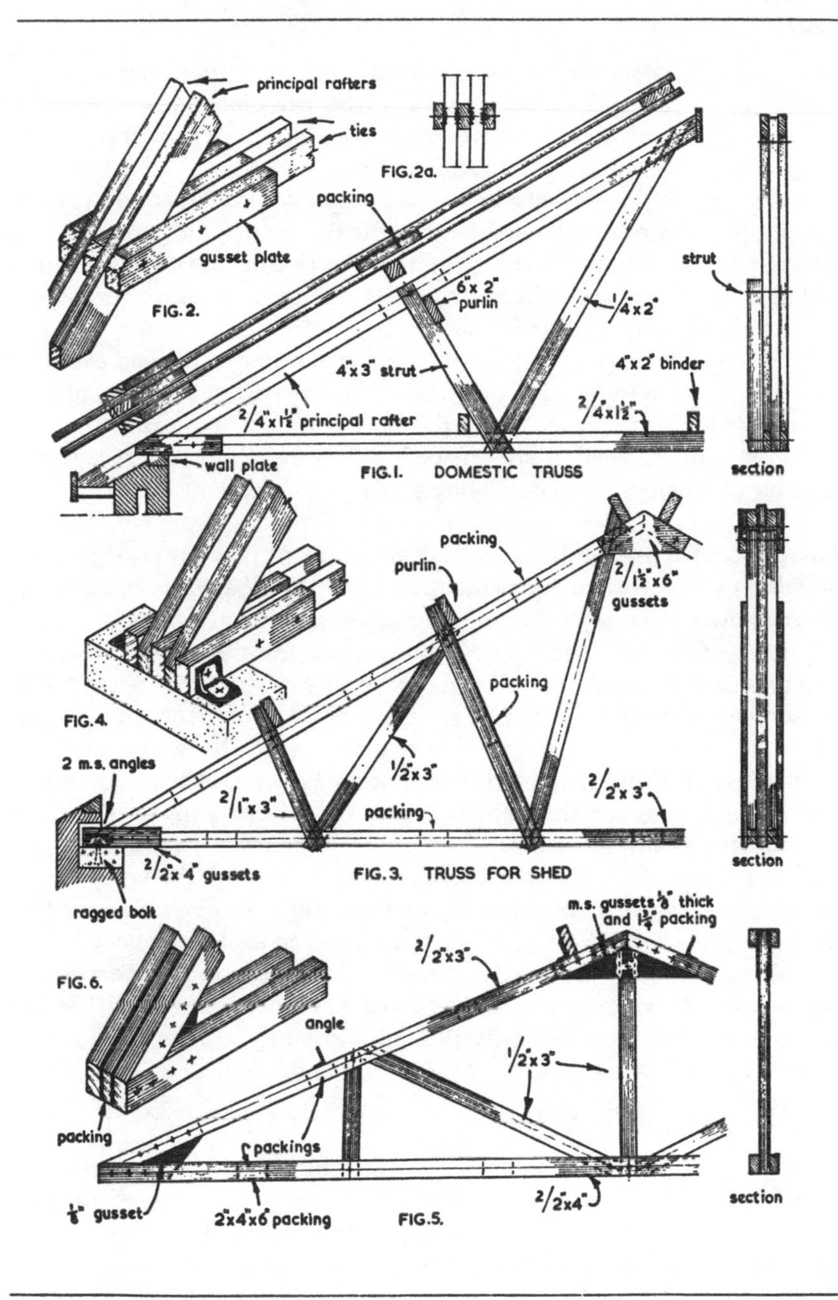
principal rafters
ties
FIG. 2a.
packing
gusset plate
FIG. 2.
6" x 2" purlin
strut
1/4" x 2"
4" x 3" strut
4" x 2" binder
2/4" x 1½" principal rafter
2/4" x 1½"
wall plate
FIG. 1. DOMESTIC TRUSS
section
packing
purlin
2/1½" x 6" gussets
packing
FIG. 4.
2 m.s. angles
1/2" x 3"
2/1" x 3"
2/2" x 3"
packing
2/2" x 4" gussets
FIG. 3. TRUSS FOR SHED
section
ragged bolt
m.s. gussets ⅛" thick and 1¾" packing
2/2" x 3"
FIG. 6.
angle
1/2" x 3"
packing
packings
⅛" gusset
2" x 4" x 6" packing
FIG. 5.
2/2" x 4"
section

One 4 in. by 2 in. tie is used in each half, and these are situated between the pairs of timbers forming the principal rafters and the lower chord. They extend down from the ridge to where the foot of the purlin support meets the lower chord.

Note that centre lines are important when setting out these trusses as they give the positions of the holes for the bolts which should pass through the centres of all the timbers. The purlins are 6 in. by 2 in., and their supports are 4 in. by 3 in. bolted on an outside surface of the principal rafter and lower chord.

When positioning on the roof, the principal rafter is notched over the wall plate, as seen in the drawing. The trusses should be placed approximately 6 ft. apart along the length of the roof, and binders, 4 in. by 2 in. in section should be used as required to prevent the sagging of the intermediate ceiling joists.

Large shed truss. Fig. 3 is the elevation of a truss for a large shed on which a covering such as corrugated iron or asbestos is to be used. That shown is suitable for a building with a span of about 25 ft. As the roof covering is much lighter than for the domestic truss, where it is presumed tiles will be used, the timbers for this truss can be somewhat smaller in size. For instance, the lower chord and the principal rafters each comprise two pieces 3 in. by 2 in., and are 2 in. apart to allow the single ties to pass between them. The other members in this case are made up from two pieces 3 in. by 1 in. and each is on an outside surface of the main timbers of the truss.

Fig. 4 shows the joint where the principal rafter and the lower chord meet. Assuming that the walls of the building are to be brick, a plate must be placed on the top of the wall at each end, and bevelled to the pitch of the roof so that the lower ends of the roof covering can be secured. In this case concrete slabs have been provided for the trusses, and they have been fixed to the masonry with two mild steel brackets. All joints are secured with bolts and timber connectors as before.

Low-pitched roof (2). Fig. 5 is the elevation of another low-pitched roof, designed to be used with roof sheeting such as corrugated asbestos, but where angle-iron purlin supports have been fixed to the top surfaces of the principal rafters. Metal gusset plates have also been used for joining the timbers at the eaves and ridge. The gusset

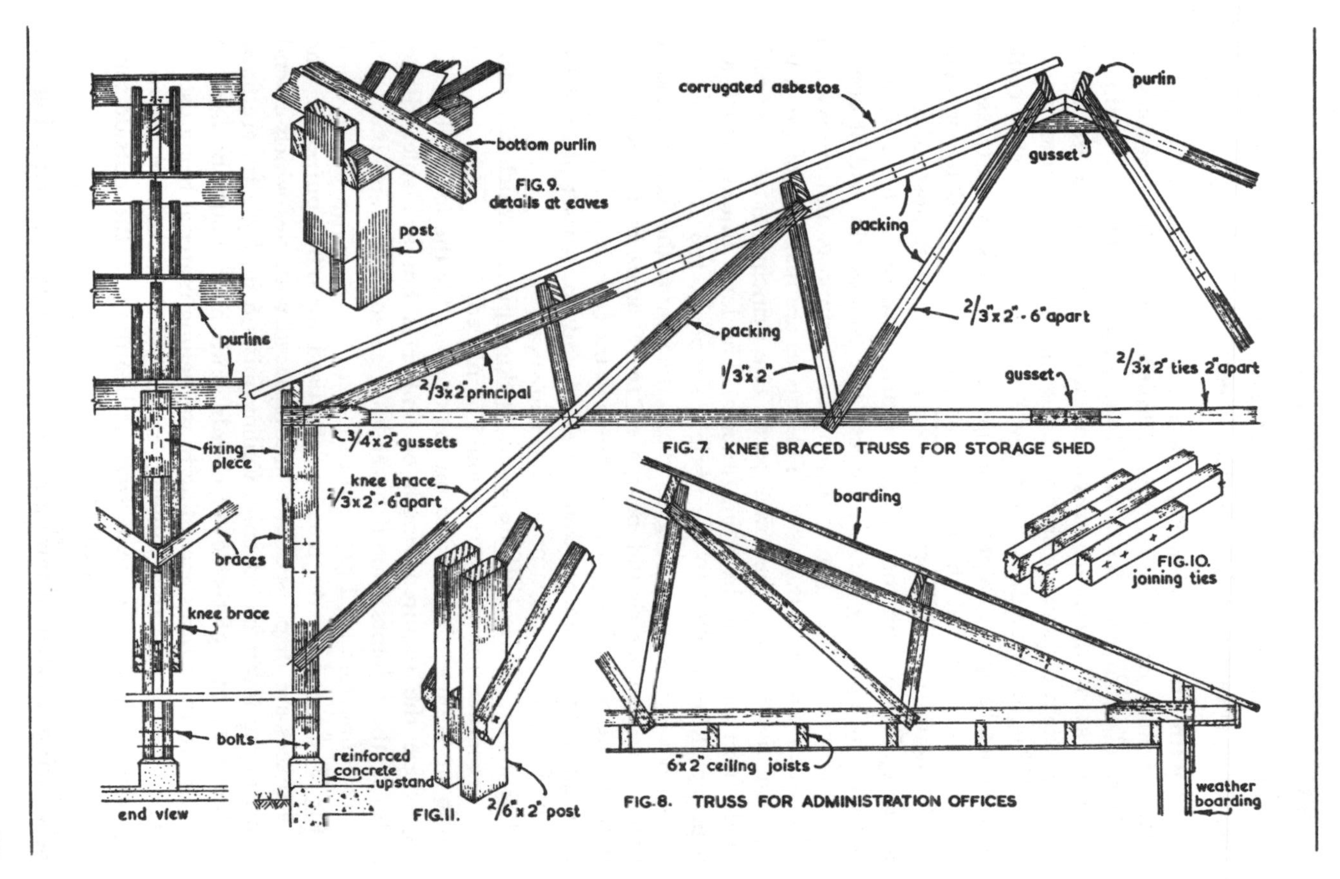

FIG. 7. KNEE BRACED TRUSS FOR STORAGE SHED

FIG. 8. TRUSS FOR ADMINISTRATION OFFICES

plates ($\frac{1}{8}$ in. thick mild steel) are in pairs, and are situated between the principal rafters and the packing pieces.

Truss with knee braces. Fig. 7 is a timber truss suitable for either a storage shed, or for the roof over administration offices, or for some other similar purpose. When knee braces are incorporated in this, it is known as a knee-brace truss, and these components are almost essential when the storage shed is of the open variety. In other words, the roof might cover the space below but the sides of the shed could be completely open except for the upright supports to the trusses.

Let us assume that the trusses are going to be used for an open shed. The upright supports could rest on a concrete upstand as seen in the end view, and they could comprise two pieces 6 in. by 2 in. spaced 2 in. apart by packing pieces. The trusses, the end of one of which is seen in Fig. 9, rest on top of the supports and are secured with a fixing piece at the eaves. The trusses are made up of two 3 in. by 2 in. ties forming the lower cord. The principal rafters are each made from two 3 in. by 2 in. timbers. The knee braces, which are 6 in. apart, are bolted to the upright supports, the chords and the principal rafter as shown. Fig. 11 shows the joint at the feet of the knee brace. The trusses should be spaced 10 ft. apart and braces fixed between the vertical supports, as shown in the end view. Three 4 in. by 2 in. gusset plates are used at the eaves and one gusset plate at the ridge. The purlins are supported as shown.

The drawings also indicate where packing pieces, which considerably strengthen the structure, should be placed.

Fig. 10 shows how to build up the timbers for the lower chord or tie, where the required length of timber is not available. Gusset plates at these joints should be at least 24 in. long, and bolts and timber connectors used between each pair.

Fig. 8 shows how the truss can be used for office purposes. The knee braces are dispensed with and replaced with ties which extend down to the lower chord only. Ceiling joists 6 in. by 2 in. are fixed to the undersides of the lower chords, and these should be fixed at 16 in. to 18 in. centres according to the ceiling material being used.

The sides of the office building can be filled in between the truss supports with studding and weather-boarding to the outside, and plaster boarding to the inside surfaces.

Bow-string truss. The new form of bow-string truss is seen in Fig. 12.

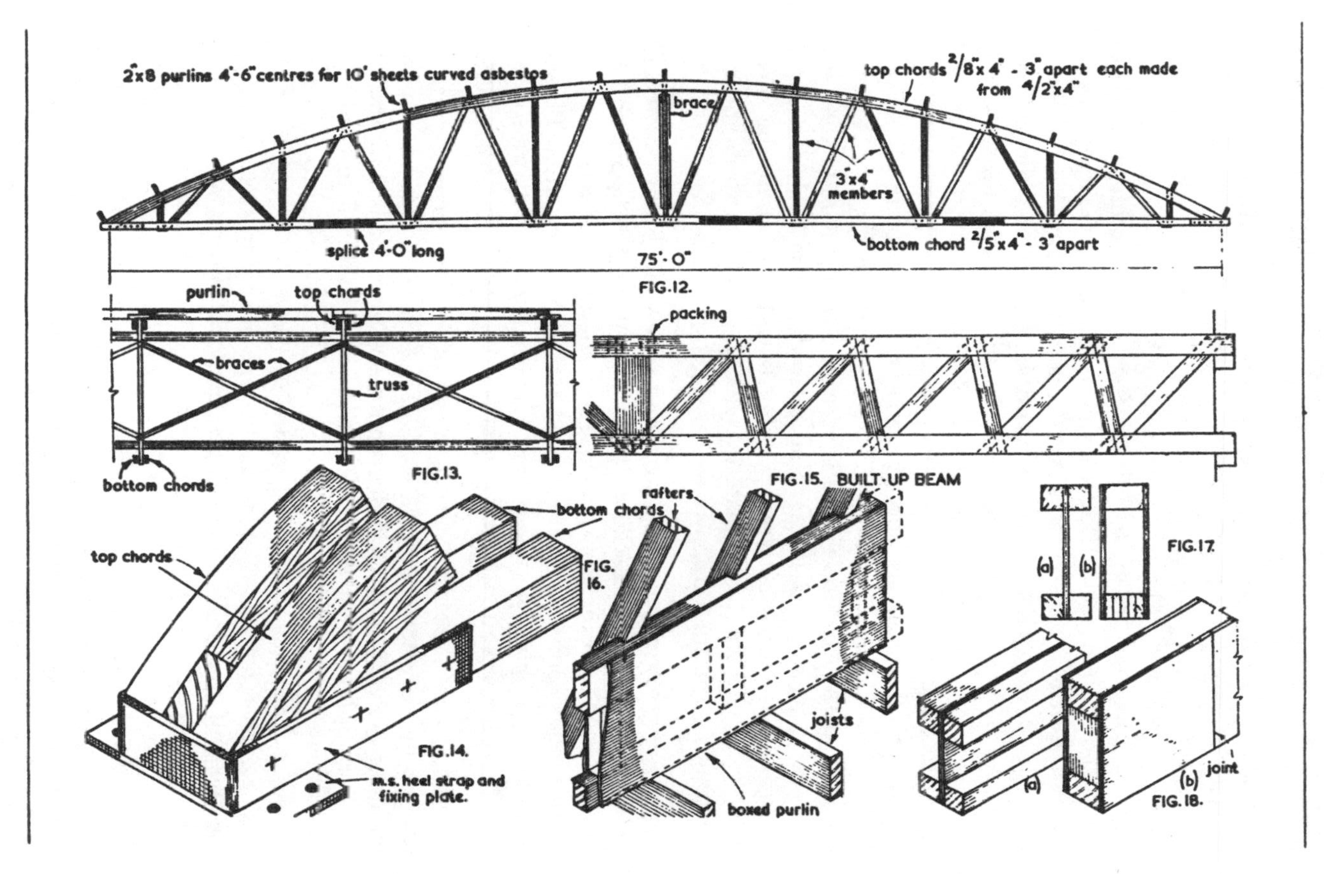

FIG. 12.

FIG. 13.

FIG. 14.

FIG. 15. BUILT-UP BEAM

FIG. 16.

FIG. 17.

FIG. 18.

Although this has been designed to span 75 ft., these trusses are capable of spanning a much greater distance, 150 ft. and more. Trusses comprise two top chords and two lower chords with purlin supports between. The top chords each have overall dimensions of 8 in. by 4 in., and are made up from four 4 in. by 2 in. timbers built up to the desired curve. Bottom chords can be of solid timber but can also be built up from smaller pieces in the same way as the top chord, if desired. Each bottom chord has an overall dimension of 5 in. by 4 in. The purlin supports are all 4 in. by 3 in. and are positioned as shown in Fig. 12. Lower chords should be given a camber of approximately 3 in.

Bow-string trusses are spaced along the length of the building at 12 ft. to 14 ft. centres and well braced, as seen in Fig. 13. A view of the joint at the eaves of a truss is seen in Fig. 14. A mild steel strap is made to pass round the outside surfaces of the heel, and a 3 in. packing piece placed between the chords will keep these the correct distance apart. The fixing plate is welded to the heel strap and used for securing the truss in position.

Truss for roof decking. Fig. 15 shows a view of a beam which could be used to carry roof decking. This consists of two upper and two lower chords with the pieces between positioned so that one bolt and three timber connectors are required for each intersection. Plywood box beams and plywood web beams, Figs. 16, 17 and 18, are also used a great deal for structural work in schools and buildings of this sort.

Open-type trusses. Figs. 19 and 22 illustrate open-type timber trusses, built on traditional lines. These trusses are found in public buildings such as church halls, churches, town halls and the like. They exemplify why so much research has gone into the design of roof trusses since the war. Large timbers, such as those shown in these trusses, are most uneconomical in present times.

The open-type truss, Fig. 19, is the type of roof truss seen supporting the roof to a village hall with a span of, say, up to 30 ft. Each truss consists of two principal rafters, a collar, and two wall pieces. The wall pieces, which help to prevent the feet of the principal rafters from spreading, are bolted to the walls of the building and rest on stone corbels. At the top, they are bridle-jointed to the lower ends of the principal rafters. To help restrict the tendency of the feet of the principal rafters to spread, mild steel straps are bolted to the

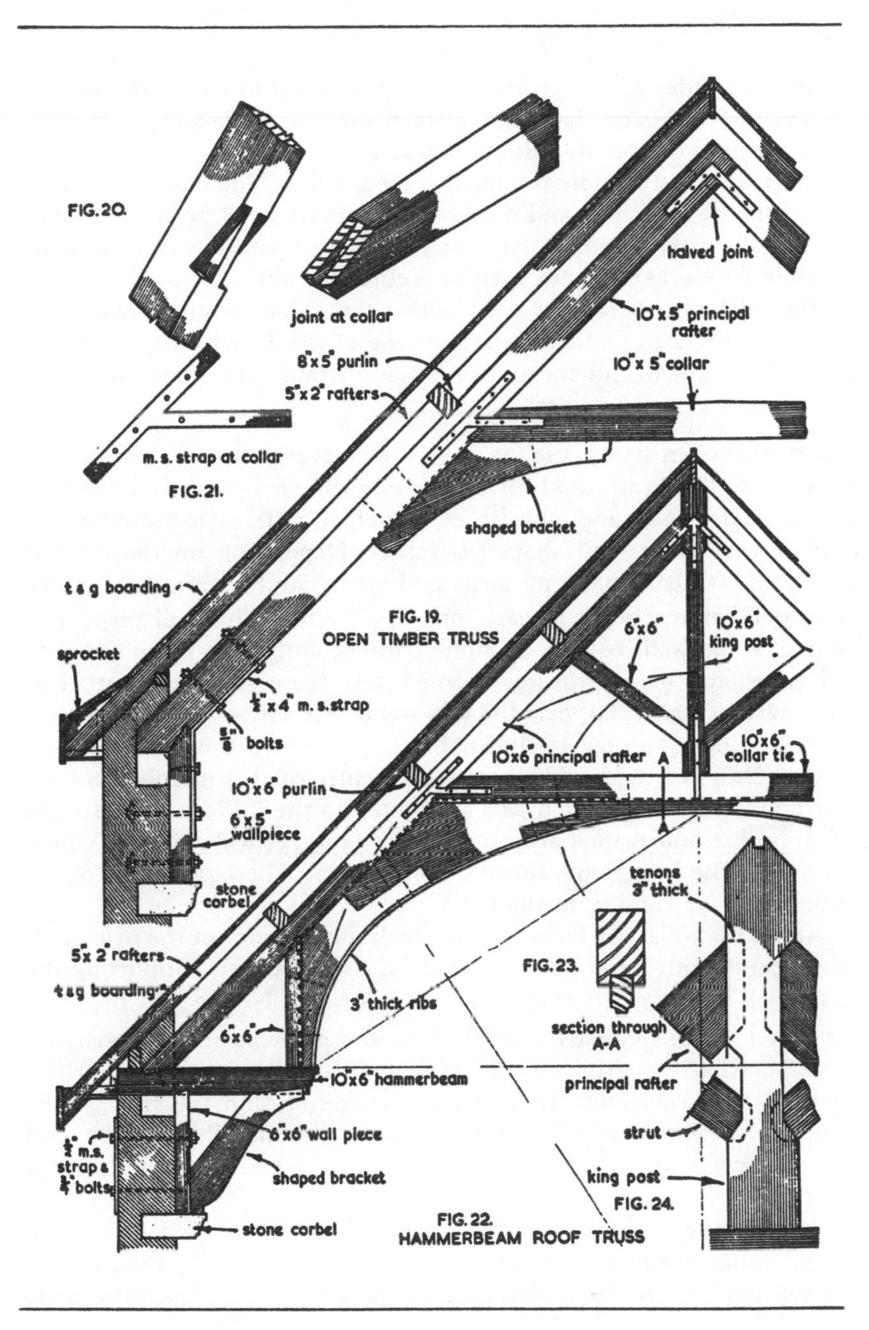
FIG. 20.
joint at collar
halved joint
10"x 5" principal rafter
8"x 5" purlin
5"x 2" rafters
10"x 5" collar
m. s. strap at collar
FIG. 21.
shaped bracket
t & g boarding
FIG. 19.
OPEN TIMBER TRUSS
6"x 6"
10"x 6" king post
sprocket
½"x 4" m. s. strap
⅝" bolts
10"x 6" principal rafter
A
10"x 6" collar tie
10"x 6" purlin
6"x 5" wallpiece
stone corbel
tenons 3" thick
5"x 2" rafters
t & g boarding
FIG. 23.
3" thick ribs
section through A-A
6"x 6"
10"x 6" hammerbeam
principal rafter
6"x 6" wall piece
strut
½" m.s. strap & ¾" bolts
shaped bracket
king post
FIG. 24.
stone corbel
FIG. 22.
HAMMERBEAM ROOF TRUSS

underneath sides of the principals and the fronts of the wall pieces.

A purlin is situated halfway between the wall plate supporting the common rafters and the ridge board, and the principal rafters are supported directly below the purlins by a collar. The collar is bridle-jointed to the principal, and mild steel straps are used, front and back, to strengthen these joints. Top ends of the principals are halved and, again, mild steel straps are used to secure this joint.

Shaped brackets can be used below the collar to give rigidity to the truss. Figs. 20 and 21 show the type of bridle joint to use at the ends of the collar and the purpose-made straps which are bolted to the timbers at these points.

Hammerbeam truss. Fig. 22 shows the elevation of a hammerbeam roof truss. These are used for supporting the roofs over buildings of up to 45 ft. span and consist of principal rafters, hammerbeams, wall pieces, collar, and shaped brackets. Depending on the design, some trusses also have king posts and struts, and occasionally, more than one hammerbeam to each principal rafter. The wall pieces are bolted to the wall, resting on stone corbels, and are bridle-jointed to the horizontal hammerbeam. A bracket rests on the same corbel as the wall piece and supports the other end of the hammerbeam.

The principal rafter is shown bridle-jointed to the hammerbeam over the wall of the building, and the centre of this member is supported by a collar. Shaped ribs are bolted to the underneath surfaces of the collar and principal rafters, and to the vertical member which rises from the inside end of the hammerbeam. These, as in the open-type roof, give rigidity to the truss.

Above the collar is a king post to which the top ends of the principals are secured with struts from the foot of this member, supporting the principals halfway between the collar and ridge. Notice that the truss has three purlins equally spaced along the principals, and assistance in supporting the principals at these points is given by the vertical piece at the lower end, the collar at the centre, and the strut at the top. Fig. 23 shows a section through the collar and shaped ribs, and Fig. 24 gives alternative methods for jointing members at the king post.

As both the trusses shown on page 41 are of the open variety, in other words all timbers of the roofs can be seen from the floor below, all the timbers in the roof should be prepared and should be good quality material. The roof is usually boarded with tongued-and-grooved boarding, face side downwards, and all timbers should be chamfered or moulded where possible.

7 Geometry and the Steel Square in Roofing

Despite what some craftsmen say regarding the use of the steel square in roofing, a good knowledge of geometry is still necessary to enable one to become efficient in the use of this valuable tool. The steel square is a geometrical instrument, and, so far as roofing is concerned, it is invaluable because, having obtained the span and pitch of the roof, it enables the lengths and the bevels of a roof to be ascertained. This also applies, of course, to the man who uses geometry and drawing instruments. Having obtained the span and the pitch, he should then be able to develop the lengths and bevels of all the parts which go to make the roof. As the geometry of roofs is basically what one has to use when applying the steel square, it is good policy to go over all the work involved in obtaining the lengths and bevels of all the members of an oblique-ended roof.

Roof with oblique end. Fig. 2 is the plan of a roof with an oblique end. Note that the jack rafters at the oblique end are all at right angles to the eaves. At the top of the page is a vertical section through the roof, and this is pitched at 45°. Below the section is the plan of the roof with the oblique end at the top, Fig. 1.

Common rafter length and bevels are obtained from the vertical section. Remember that this length e–b represents the length of the rafter down to the top outside edge of the wallplate. The overhang of the eaves has to be added on to this length, and half the thickness of the ridge board will have to be deducted, see Figs. 6 and 7. Now turn to the hip a–e at the lower end of the plan. To obtain its length construct a right angle at e and make e–g equal to the rise of the common rafters e′–e, see section. a–g is the length of the hips a–e and b–e. Half the thickness of the ridge board must be deducted from this length, see Figs. 8 and 9. The overhang will have to be added to the length obtained. The plumb bevel or cut is seen at g and the seat cut at a.

To obtain the splay cut for the two hips at the square end, construct a right angle at a so that h is on the extended centre line of the roof. Then with compass point in a and radius a–g describe an arc to give g′ on the extended line a–e. The splay bevel is a–g′–h. The backing bevel for these two hips is found by drawing the line 1–3–2 across and at right angles to the plan of the hip. With compass point in 3 and pencil set just wide enough to touch the true length of the hip a–g, describe an arc to give point 4 on a–e. The backing bevel required is 3–4–2. The lengths and bevels for the short and long hip at the oblique end of the roof are found in exactly the same way as those for the hip just covered. For the remaining bevels we should now turn to Fig. 3.

Splay cuts. To obtain the splay cuts for the jack rafters (the plumb and seat cuts are the same as for the common rafters) one or more of the roof surfaces have to be developed. To obtain the splay cut for the jack rafters which intersect with the hips at the square end, either the end of the roof a–b–e can be developed, or one of the sides adjacent to the square end. Let us choose the surface e–b–c–f. To develop this surface place the compass point in e at the top of the vertical section, and with radius e–b describe an arc to give point b′ on the horizontal line brought out from e. Drop a vertical line from b′ to give points b′ and c′ on horizontal lines brought out from b and c in the plan. e–b′–c′–f is the developed surface and the splay cut is seen near point b′. This is obtained by drawing any line which is at right angles to the eaves to pass across the developed surface.

To obtain the splay cuts to the jack rafters at the oblique end of the roof the oblique end f–c–d must be developed.

Purlins. The next bevels to develop are those for the purlins. First draw a section of the purlin, any size, as seen in the vertical section, so that the extended edge 1–2 will intersect with the centre line in e′ (Fig. 3). Next, draw the plan of the purlin as seen in the roof plan. To obtain the cuts required the two top surfaces 1–2 and 1–3 have to be developed. Draw a horizontal line through point 1 in the section, and with centre 1 and radii 1–2 and 1–3 in turn, describe arcs to give point 2′ and 3′ on the horizontal line passing through 1. Drop vertical lines from points 2′ and 3′ to intersect with horizontal lines brought

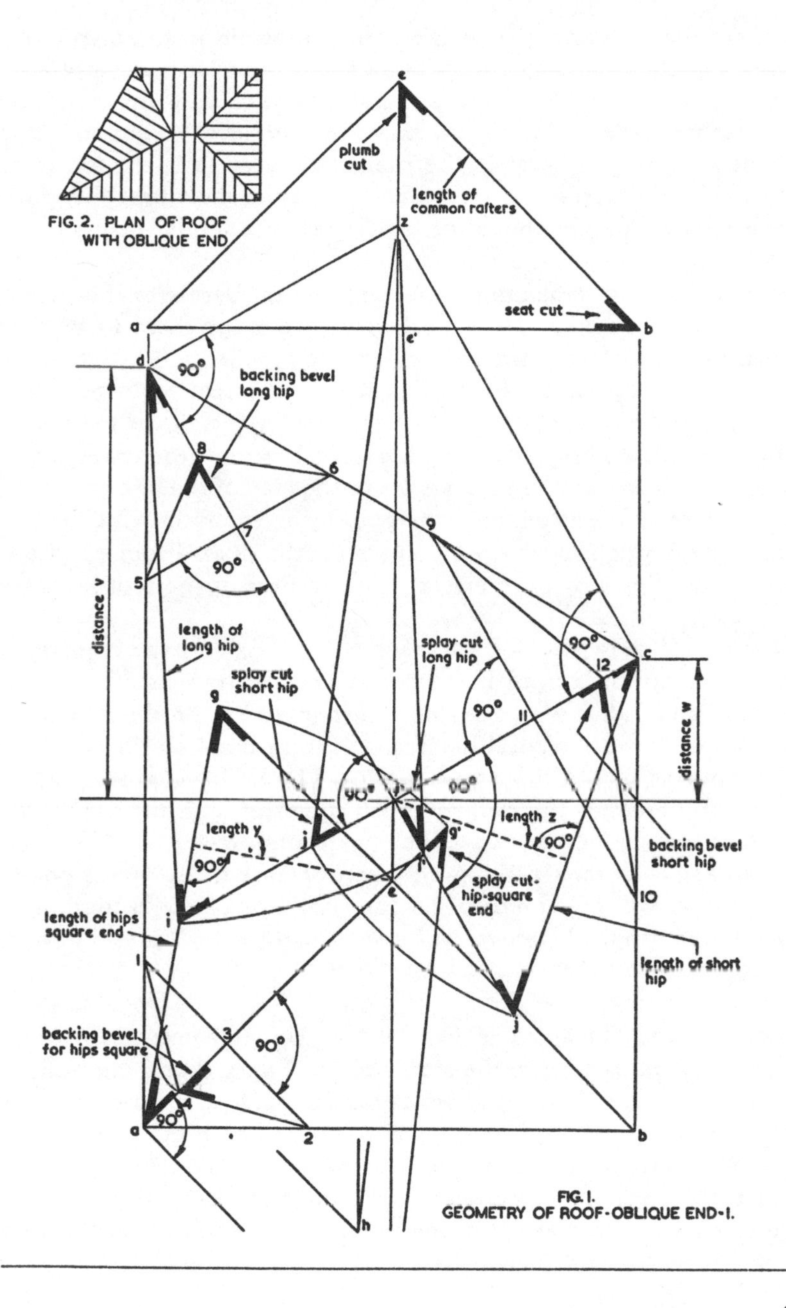

FIG. I.
GEOMETRY OF ROOF-OBLIQUE END-I.

out from points 2 and 3 at each end of the purlin in the plan to give points 2′ and 3′ at each end. The developed surfaces are 1–2′–2′–1 and 1–3′–3′–1, and the bevels are seen at the ends of these surfaces.

Those near the bottom of the page are the bevels required for the purlins at the square end of the roof, and those near the top of the plan are for the purlins which intersect with the short hip. The purlin bevels to the long hip should be developed in a similar manner.

Lip bevels. The remaining bevels are the lip bevels to the purlins, and these are applied to the purlins in order for them to be fitted underneath the lower surfaces of the hip rafters, see Fig. 4. To obtain the lip bevels to the purlins which intersect with the hips at the square end extend the edge of the purlin 1–2 down to intersect with the point e′. Next draw the horizontal trace of the lower edge of hip e–b. This is drawn at right angles to the plan of the hip, and should extend from the centre line of the plan up to intersect with the horizontal line brought over from point e on the plan. From e″ draw a vertical line upwards to intersect with the base of the roof section in e‴.

From e‴ draw the vertical trace e‴–e. Extend the edge of the purlin 2–1 upwards until it meets the vertical trace in X, and with centre e′ and radius e′–x describe an arc to give x′ on the base line of the section. Drop a vertical line from x′ to give x″ on the e–e″ line, and from this point draw a line to x‴. The lip bevel is seen at x‴.

The lip bevel for the purlins which intersect with the small hip is found in exactly the same way. The horizontal trace f″–y‴ is drawn at right angles to the small hip, the vertical trace f‴–e giving point y. Radius e′–y will give point y′ on the base line of the section, and a vertical line dropped from y′ will give y″ on the f–f″ line. Draw line y″–y‴ for the bevel.

Steel square. The above is, briefly, the geometry needed for roofing work. Now let us turn to the steel square. Fig. 5 shows one side of a Stanley steel square, and the tables on the blade are those we should be interested in when dealing with roofing. These tables, of course, deal only with rectangular roofs. The top line of the tables deals with the length of common rafters per foot run. Let us just take one set of figures to illustrate the use of the tables, the figures under the 12 in. mark on the edge of the square.

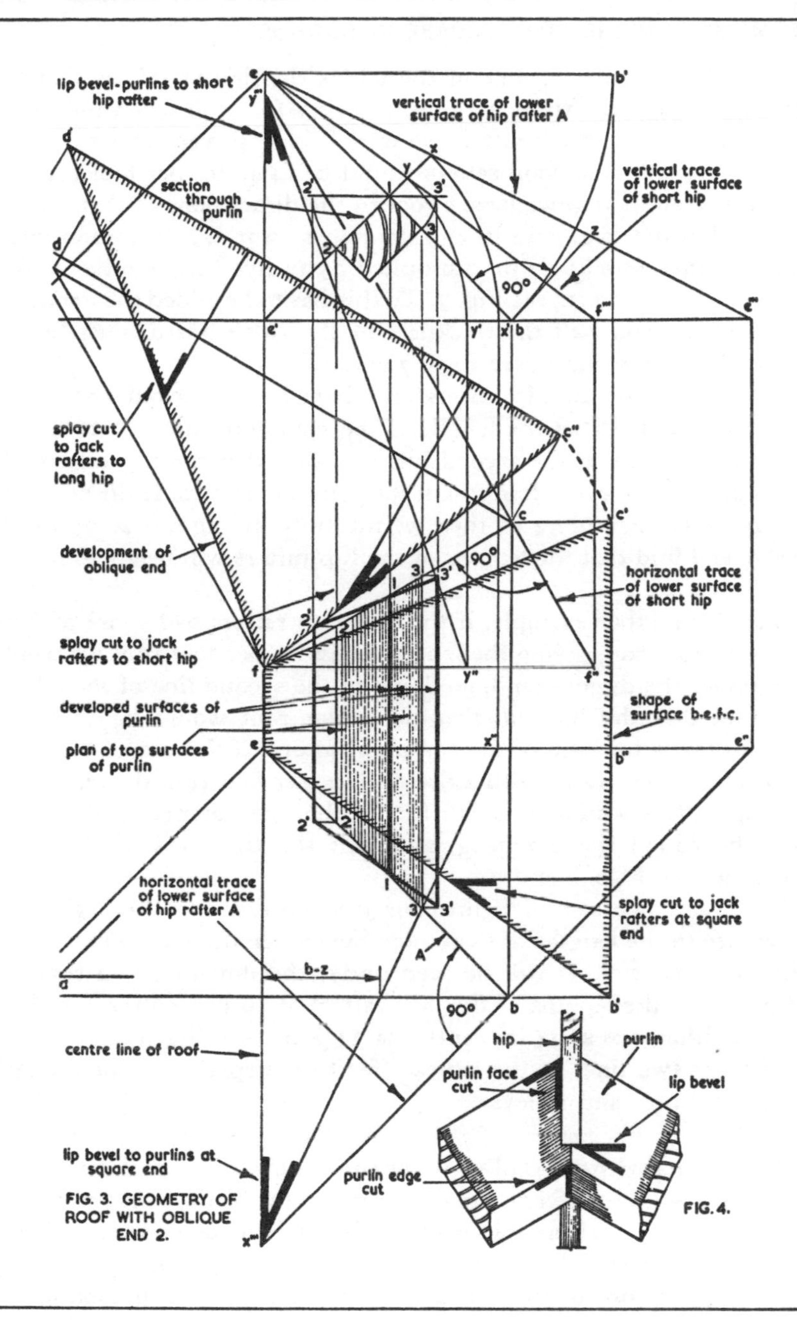

FIG. 3. GEOMETRY OF ROOF WITH OBLIQUE END 2.

If we look at the top line of the tables the figures tell us that the common rafters are 16·97 in. long for every foot run when the rise is 12 in., see Fig. 6. If the rafter rise were 8 in. for every foot run, then the length of the common rafters would be 14·42 in. per foot run.

Let us assume that we have a roof to the dimensions of that seen in Fig. 6. Half the span is 12 ft. and the rise is 12 ft. The common rafter length is therefore 16·97 in. multiplied by twelve (the number of feet run) which is 16·97 ft., see Fig. 7. To this has to be added the overhang of the eaves, and half the thickness of the ridge board must be deducted from the length, see Figs. 7 and 12.

The lengths of the hip rafters are given in the second line of the tables. If we find that the length of the common rafters is 16·97 per foot run, we look at the second line and find that the hip rafters are 20·78 in. long for every foot run of the common rafters. If the common rafters have a run of 12 ft. then we multiply the figures 20·78 in. by twelve and find that the lengths of the hip rafters will be 20·78 ft., see Fig. 8.

To take another example, if the common rafters had a rise of 8 in. for every foot run, giving the length 14·42 in. per foot run, we would look under the dimension 8, go down to the second line of the tables, and find that the hips on that particular roof would be 18·76 in. long for every foot run of the common rafters. If the run of the common rafters were 12 ft. with a rise of 8 in. per foot run, the lengths of the hip rafters would be 18·76 ft. To this measurement, of course, must be added the overhang, and half the thickness of the ridge deducted, see Figs. 8 and 9.

There are two sets of figures for jack rafters, and these give the difference in the lengths of these members when used at 16 in. centres and 24 in. centres. It will be seen under the dimension 12 that the difference in the lengths of the jack rafters at 16 in. centres is $22\frac{5}{8}$ in. and the difference at 24 in. centres is $33\frac{15}{16}$ in., see Fig. 10.

The last two lines of the tables refer to the splay cuts for the jack rafters and hips and valleys.

Calculations without tables. If one prefers to work out all the lengths and bevels without referring to the tables on the blade, or if a roof with an oblique end such as that seen in Fig. 2 is to be constructed, the following notes and diagrams will help. They are useful to those who are not quite sure of the full use of the steel square in roofing. As

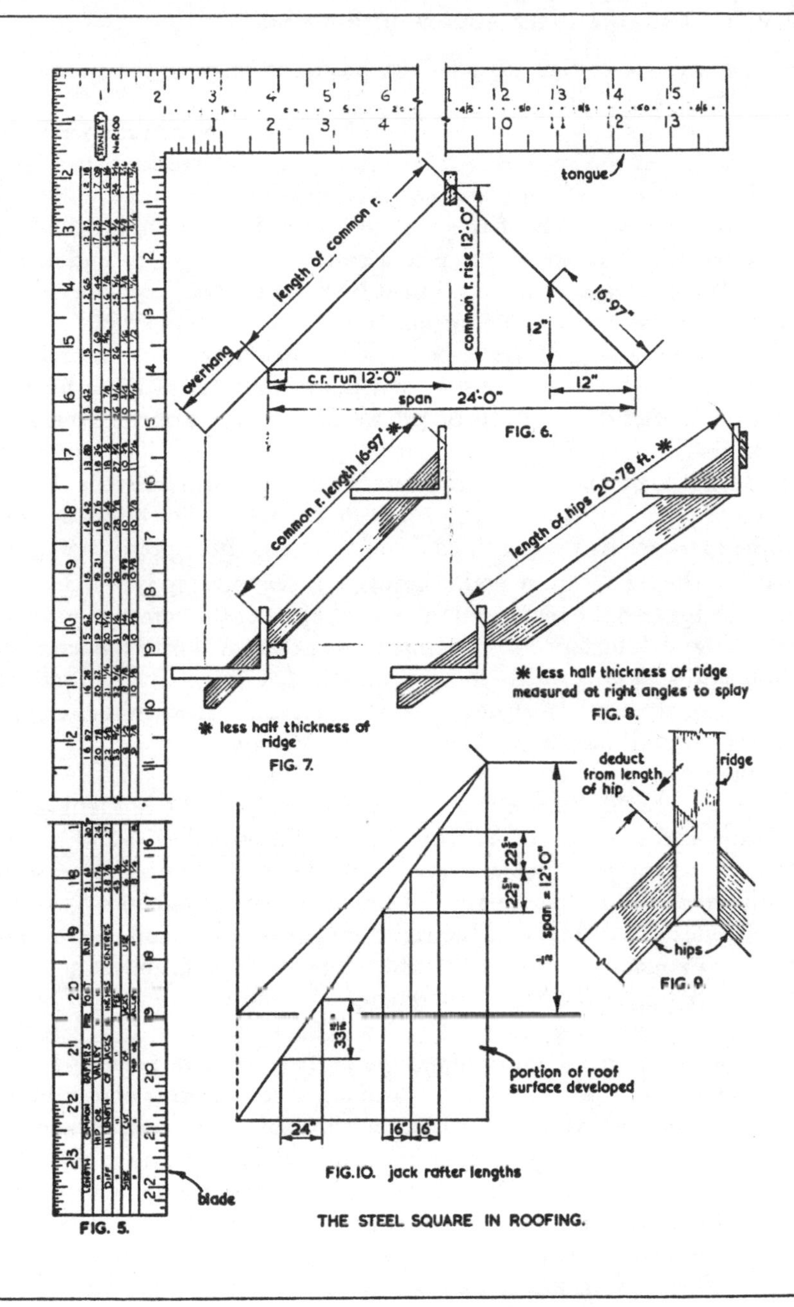

THE STEEL SQUARE IN ROOFING.

the diagrams are explained it will be useful to refer to the geometry on the foregoing pages.

Fig. 11 shows how the lengths and bevels of the common rafters are found when the span and the pitch of the roof are known. Mark half the span on the blade and set the fence of the square up at an angle equal to the pitch of the roof to obtain the length and bevels. Remember that you are working to a scale of 1 in. to 1 ft. and so all the lengths will have to be multiplied by twelve. Also, half the thickness of the ridge must be deducted from the length obtained.

Next, the overhang of the eaves must be calculated, Figs. 12 and 13. Determine the distances a and b, and mark the total of these figures on the blade. Set the fence at the pitch and the length of the overhang is obtained.

The lengths of the hip rafters must be obtained in two steps. First, the run of the hips, square end. Mark half the span on the blade and half the span on the tongue. The distance across these points will give the run of the hips, square end. Fig. 14. The second step is to mark the run of the hips on the blade and the rise of the roof (or common rafters) on the tongue. The fence fixed across these points will give the lengths, plumb, and seat cuts for the hips, square end, Fig. 15. Remember the overhang of the hips must now be calculated and added to the length. Half the ridge thickness measured at right angles to the splay must be deducted.

To find the splay cut for the hips, square end, set the hip length on the blade and the hip run on the tongue. The bevel is found at the blade end of the fence. If reference is made to Fig. 1 it will be seen that the splay cut to the square end hips has been obtained by having the hip length as one side of the right-angled triangle, and the second side a–b is equal to the run of the hip. The bevel is found at g.

The backing for the hips, square end, is found in two stages. First, the distance y has to be found, Fig. 17. This distance in Fig. 1 is the amount we have to open the compasses so that they just touch the true length of the hip rafter. To find distance y set the run of the hips, square end, on the blade, and the rise of the roof on the tongue, and fix the fence over these two points. Measure with a two-foot rule the shortest distance between the corner of the square and the edge of the fence. This measurement is distance y. The second step is to set the fence to the run of the hip on the blade and distance y on the tongue. The backing bevel is at the tongue end of the fence, Fig. 18.

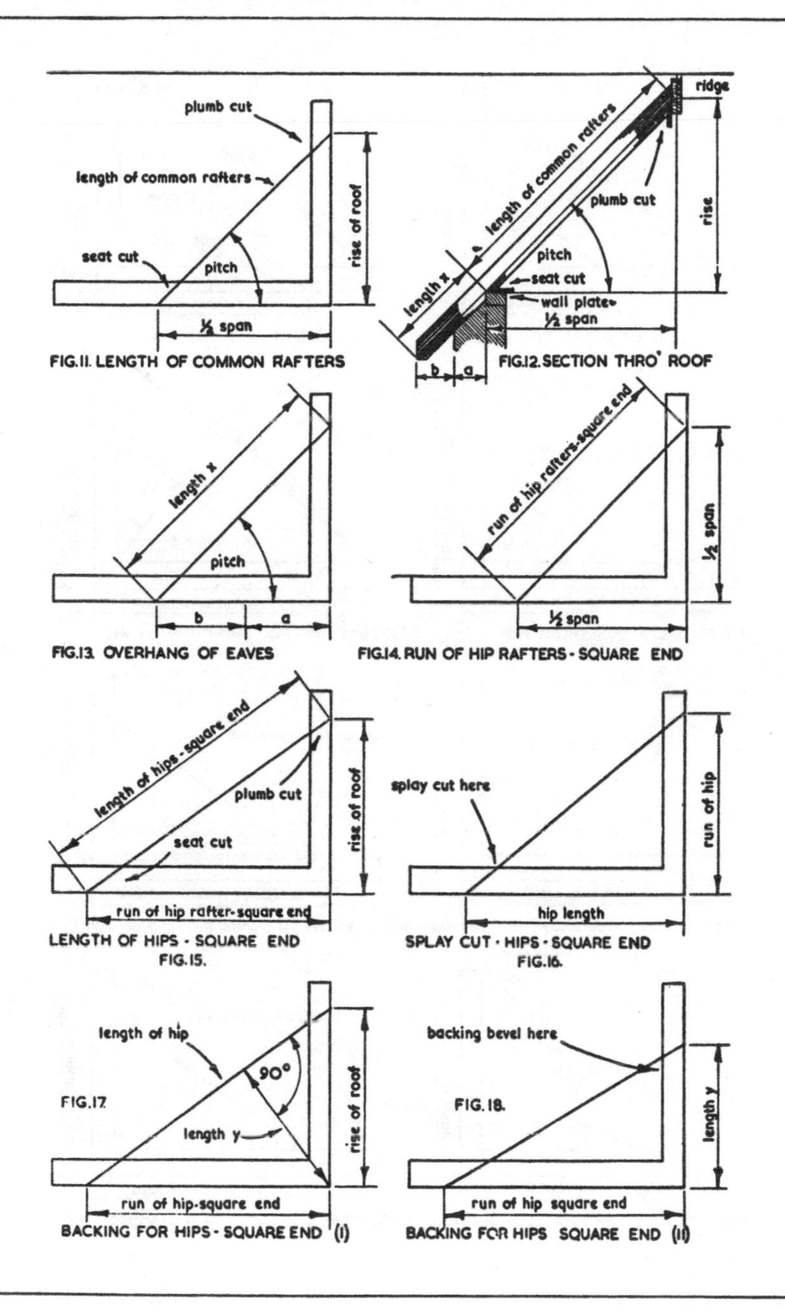
plumb cut
length of common rafters
rise of roof
seat cut
pitch
½ span
FIG.11. LENGTH OF COMMON RAFTERS
ridge
length of common rafters
plumb cut
rise
length x
pitch
seat cut
wall plate
½ span
b
a
FIG.12. SECTION THRO' ROOF
length x
pitch
b
a
FIG.13. OVERHANG OF EAVES
run of hip rafters-square end
½ span
½ span
FIG.14. RUN OF HIP RAFTERS - SQUARE END
length of hips - square end
plumb cut
seat cut
rise of roof
run of hip rafter-square end
LENGTH OF HIPS - SQUARE END
FIG.15.
splay cut here
run of hip
hip length
SPLAY CUT - HIPS - SQUARE END
FIG.16.
length of hip
90°
FIG.17.
length y
rise of roof
run of hip-square end
BACKING FOR HIPS - SQUARE END (I)
backing bevel here
FIG.18.
length y
run of hip square end
BACKING FOR HIPS SQUARE END (II)

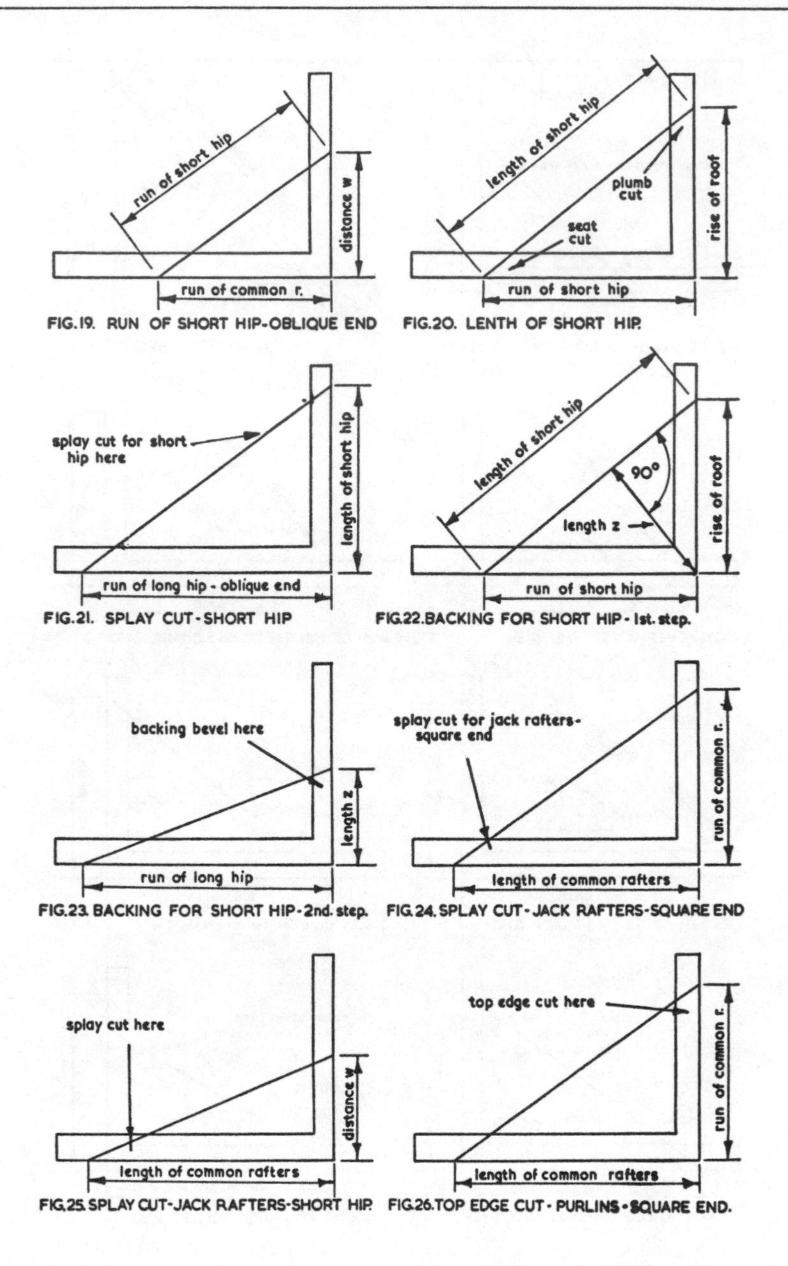
run of short hip
distance w
run of common r.
FIG.19. RUN OF SHORT HIP-OBLIQUE END
length of short hip
plumb cut
seat cut
rise of roof
run of short hip
FIG.20. LENTH OF SHORT HIP.
splay cut for short hip here
length of short hip
run of long hip - oblique end
FIG.21. SPLAY CUT-SHORT HIP
length of short hip
90°
length z
rise of roof
run of short hip
FIG.22.BACKING FOR SHORT HIP - 1st. step.
backing bevel here
length z
run of long hip
FIG.23. BACKING FOR SHORT HIP-2nd. step.
splay cut for jack rafters-square end
run of common r.
length of common rafters
FIG.24. SPLAY CUT-JACK RAFTERS-SQUARE END
splay cut here
distance w
length of common rafters
FIG.25. SPLAY CUT-JACK RAFTERS-SHORT HIP.
top edge cut here
run of common r.
length of common rafters
FIG.26.TOP EDGE CUT - PURLINS - SQUARE END.

We now come to the lengths and bevels for the hips at the oblique end. Let us take the short hip. To obtain the run of the short hip, mark half the span (or run of common rafter) on the blade, set the fence from this point at the same angle the small hip makes with the eaves at the square end of the roof (30°). Note the distance w as well as the run of the short hip, Fig. 19.

To obtain the length of the short hip, set off the run of the hip on the blade and the rise of the roof on the tongue. The fence will give the length of the hip as well as the plumb and seat cuts, Fig. 20. To obtain the splay cut for the short hip, set off the run of the long hip on the blade. This is equal in length to the line c–z, Fig. 1. Setting the length of the short hip on the tongue gives the splay cut at the tongue end of the fence, Fig. 21.

The backing for the short hip is found, as for that at the square end, in two stages. The first step is to obtain length z. This can be measured with the rule when the run of the short hip has been marked on the blade and the rise of the roof on the tongue. The second step is to mark the run of the long hip on the blade and length z on the tongue to obtain the backing bevel at the tongue end of the fence, Fig. 23.

Splay cuts. The splay cut for the jack rafters can be found simply by recalling what would be done if a surface of the roof had to be developed by geometrical means. The splay cut for all the jack rafters at the square end of the roof can be found by marking the length of the common rafters on the blade (this is equal to the width of the surfaces) and the run of the common rafters on the tongue (this is equal to b′–b″ on the plan). The splay cut is found at the blade end of the fence, Fig. 24.

The splay cut for the jack rafters adjacent to the short hip is found by marking the length of the common rafters on the blade (the width of the surface) and distance w on the tongue to get the bevel at the blade end of the fence, Fig. 25.

Purlin cuts. Obtaining the cuts for the purlins is similar to developing the surfaces of a roof. The top surface of a purlin can be taken as a small portion of the roof surface which it supports. Let us consider the purlins at the square end of the roof first. It will be seen from Fig. 3 that the top surface is a portion of the roof surface. Consequently, when developed the end of the purlin will be of similar shape as the

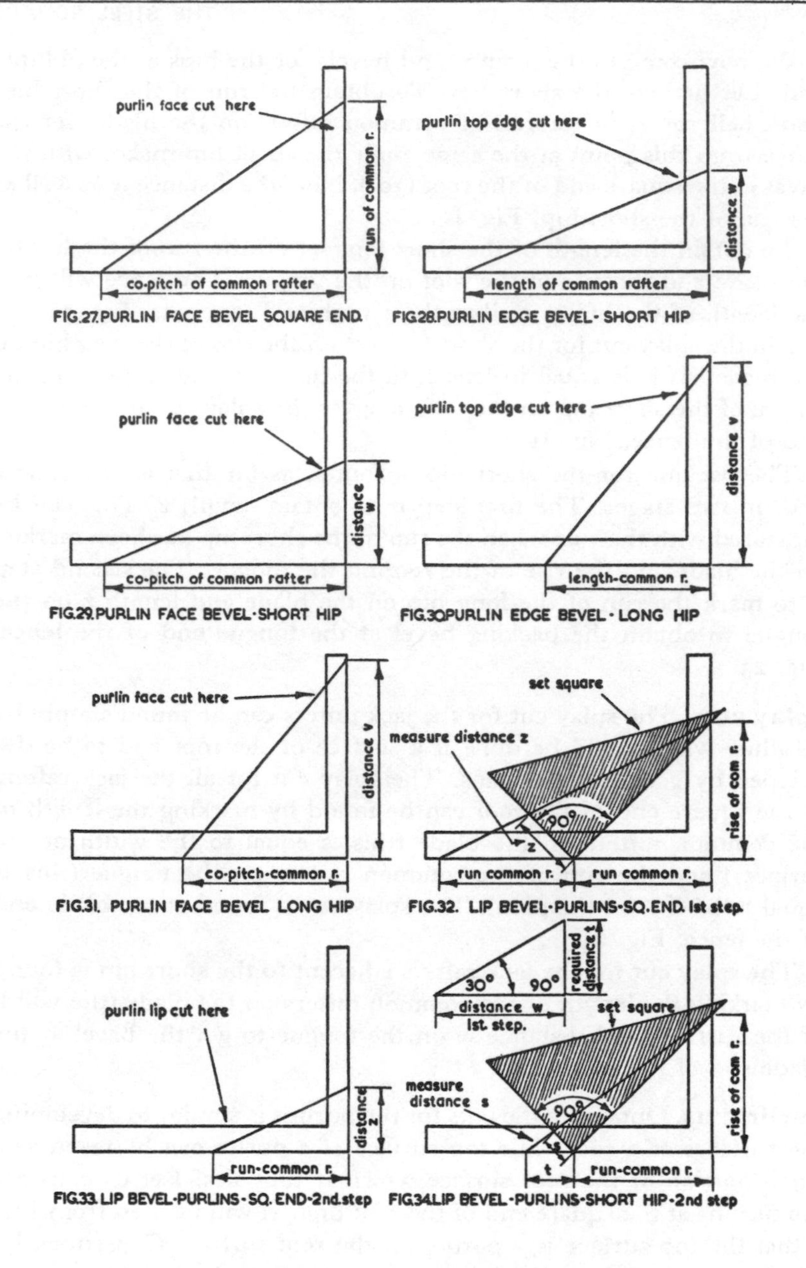

FIG.27. PURLIN FACE BEVEL SQUARE END.

FIG.28. PURLIN EDGE BEVEL - SHORT HIP

FIG.29. PURLIN FACE BEVEL - SHORT HIP

FIG.30. PURLIN EDGE BEVEL - LONG HIP

FIG.31. PURLIN FACE BEVEL LONG HIP

FIG.32. LIP BEVEL-PURLINS-SQ. END. 1st step.

FIG.33. LIP BEVEL-PURLINS-SQ. END-2nd. step

FIG.34. LIP BEVEL-PURLINS-SHORT HIP-2nd step

end of the roof surface. Fig. 26 should make it quite clear that when developing the splay cut for the jack rafters the top edge bevel for the purlin can be obtained from the other end of the fence. When dealing with roofs pitched at 45° the side bevel for the purlin is exactly the same as for the top edge where the roof is rectangular.

This however only applies to roofs of 45°. If the pitch is more or less than 45° the bevel for the side is different from the top edge. So as not to make a common error, it is always best to follow a rule when developing the bevels to purlins. The length of the common rafter should be used when developing the top edge cut of the purlin, and the co-pitch (see Fig. 36 a and b) used when developing the side cut. To develop the top edge cut for the purlins adjacent to the short hip at the oblique end, mark off the length of the common rafter on the blade and distance w on the tongue (Fig. 1). The bevel required is at the tongue end of the fence, Fig. 28.

To obtain the face bevel to the purlins at the short hip, mark the co-pitch on the blade and the distance w on the tongue to obtain the bevel at the tongue end of the fence, Fig. 29.

Hips. Now let us turn to the long hip at the oblique end. Mark the length of the common rafter on the blade and the distance v, Fig. 1, on the tongue. The top edge bevel is found at the tongue end of the fence, Fig. 30. The face bevel to the long hip is found by marking the co-pitch of the rafters on the blade and the distance v on the tongue to obtain the bevel at the tongue end of the fence, Fig. 31.

Co-pitch. One might well ask, 'how does one find the co-pitch of a roof?' Let us take a look at Fig. 36b. The pitch of the roof is 30°, which means the co-pitch angle is 60°. The co-pitch is always at right angles to the slope of the roof surfaces. To find the co-pitch of the 30° roof, Fig. 36b, mark off half the span on the blade, set the fence up to 60° from the half-span mark, and measure with a rule the distance along the fence giving the length of the co-pitch.

Purlin lip bevels. The only remaining cuts required are the purlin lip bevels. Study Figs. 3 and 32 together. The horizontal trace of the hip b–e terminates at e''. The line e–e'' is equal to twice the run of the common rafters. Mark this distance on the blade. The vertical trace of the same hip is e'''–e. The distance between e and e' is equal to the rise of the roof. Mark the rise on the tongue and fix

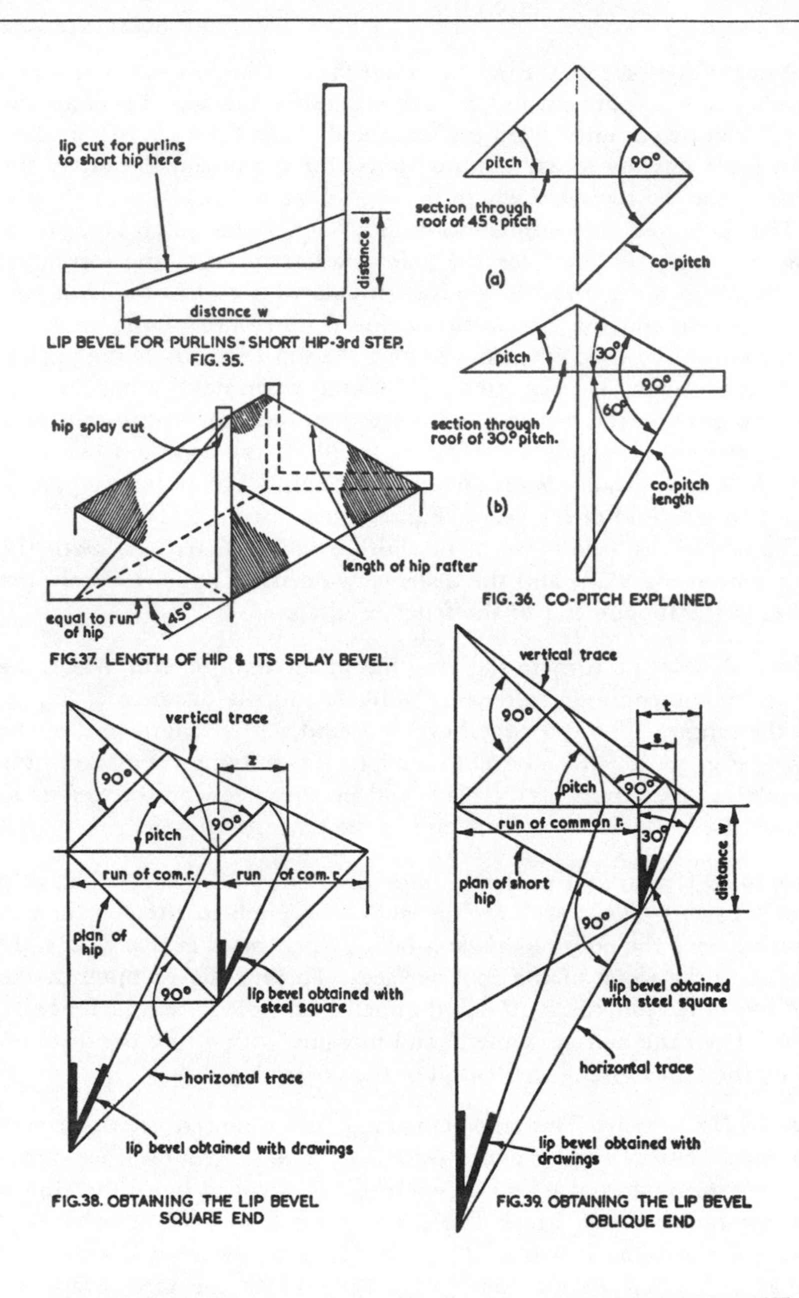

LIP BEVEL FOR PURLINS - SHORT HIP - 3rd STEP. FIG. 35.

FIG. 36. CO-PITCH EXPLAINED.

FIG. 37. LENGTH OF HIP & ITS SPLAY BEVEL.

FIG. 38. OBTAINING THE LIP BEVEL SQUARE END

FIG. 39. OBTAINING THE LIP BEVEL OBLIQUE END

the fence over these two points. Now place a set square over the steel square as seen in the drawing, and measure distance z. This is equal to the distance b–z in the section and b–z in the plan. The next step in obtaining the lip bevel for the square end purlins is to mark off the run of the common rafters on the blade and distance z on the tongue. The lip bevel is at the blade end of the fence, Fig. 33 (see also the plan of the roof Fig. 4).

Obtaining the lip bevel for the short hip is illustrated in Fig. 34 (first and second steps). Fig. 35 (third step) is explained in geometrical form in Fig. 39. The reader should study these drawings and satisfy himself that the bevel obtained is the correct one.

Fig. 38 further illustrates how the lip bevel for the square end purlins is obtained, and Fig. 37 shows how the steel square is used for developing the splay cuts for the hip rafters.

8 Timber Buildings

There are many advantages in a timber-built house. The amount of time required to build it is considerably less than that for a traditionally constructed house of similar dimensions; there is considerable fuel saving because timber has a high insulation value; with dry wall construction, plaster board, or similar covering material, no delay is required before decoration begins; and frozen pipes are almost unknown in a well-designed and well-built timber house.

Still more important, so far as the builder is concerned, factory-built methods can be adopted, and, with proper supervision, semi-skilled labour can be used on much of the work. Portable tools, too, play a big part in keeping costs down.

Foundation loads are much lighter than for the traditional brick house, and so the sizes of foundations can be reduced. Foundations are of concrete and the walls up to a point above the ground level can be either brickwork or concrete which is cast 'in situ'

Methods of construction—platform system. There are several kinds of construction used for timber houses, amongst them being the platform system. Fig. 1 is a vertical cross-section through a house built on the platform system. The foundations are of the kind which would be provided for a brick house, but, of course, the concrete footings are considerably smaller. Their size, however, still depends on the weight of the structure to be placed on them. Brick walls 9 in. thick are shown in the drawing, and these are brought up to a height of at least 6 in. above ground level. When building the main walls ½ in. diameter bolts must be inserted in the wall every 6 ft. around the building to provide a fixing for the 4 in. by 2 in. plates on which the ground floor joists rest.

The internal walls, which are of 4½ in. brickwork, should be brought up to the same level as the outer walls. The site within the main walls of the house, having had its vegetable soil removed, should be covered with at least 4 in. of hard-core and then with a 4 in. surface of concrete. Traditional methods of constructing ground floors can be followed by building honeycombed sleeper walls across the areas within the main walls so that the 4 in. by 2 in. floor joists will have adequate support throughout their lengths. When the main, internal and sleeper walls have been brought up to correct level, damp-proof course material is laid on the top surfaces. This will ensure that no moisture will rise up through the concrete and brickwork and so attack the timber.

Plates of 4 in. by 2 in. section are now laid on the top surfaces of all the walls with the damp-proof course material between. As an added precaution against damp, air bricks or ventilators must be built into the foundation walls just below the damp-proof courses. These allow fresh air to enter the space below the ground floor timbers and so help to keep their moisture content down to a safe level.

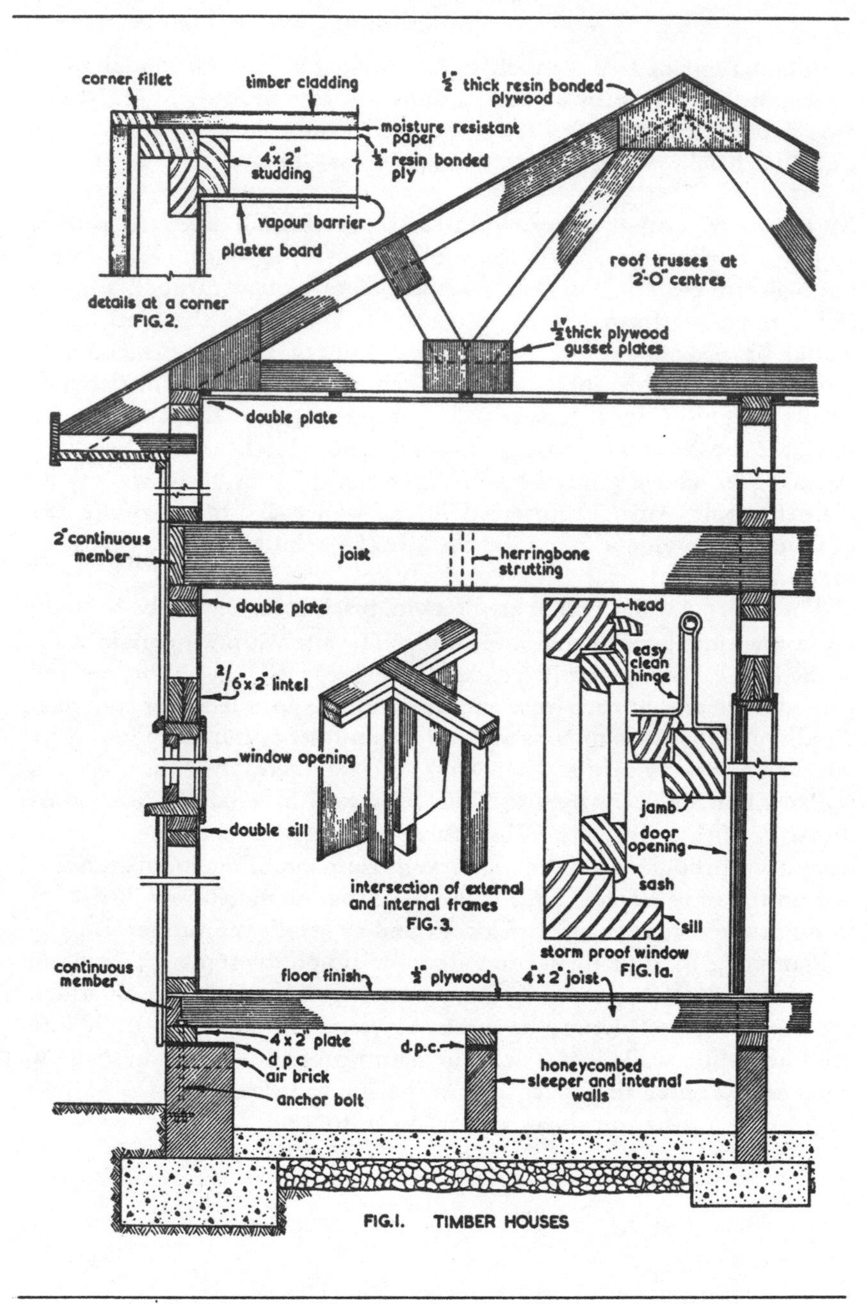

FIG. 1. TIMBER HOUSES

Ground floor platform. The next step is to build a platform (hence the name of the system) on top of the plates. This platform consists of floor joists spaced at 16 in. centres, and a continuous 4 in. by 2 in. member which makes an outer frame to the platform. When the joists have been securely nailed to the plates, and the continuous member to the joists, the surface of the platform can be laid, using ½ in. thick sheets of plywood. The latter, which are nailed to the joists and continuous member, provide a working platform from which the next operation can be carried out.

Wall frames. The manufacture of the wall frames follows, and on these the joists of the first floor rest. These frames are constructed with 4 in. by 2 in. studs at 16 in. centres, each frame having two pieces of 4 in. by 2 in. timber to form the head. The external frames can be covered on their outside surfaces with ½ in. thick sheets of resin-bonded plywood, and as each frame is completed it can be man-handled into position and held vertically by the use of braces or props.

Window and door openings are made in the frames as seen in Figs. 5 and 6. These show that the lintels above the openings are made from two pieces of 6 in. by 2 in. timbers on edge. The jambs of the openings, too, consist of two thicknesses of timber, these being 4 in. by 2 in.

There are several arrangements for the positions of the corner studs, and one of these is shown in Fig. 2. An additional stud has been placed in the corner made by the two ends of the frames to provide a fixing for the interior wall sheeting. Fig. 3 shows how the intersections between the main and internal walls are dealt with. Notice that all the frames have double plates at their tops.

Platform for upper floor. When the lower external and internal frames have been erected and secured, it is necessary to provide a platform from which the next stage can be carried out. This platform is constructed with the first floor joists resting on the top plates of the ground floor frames, a continuous member again passing round the whole perimeter of the floor. This member and the floor joists are of greater depth than the ground floor joists because the upper joists have no intermediate supports. The depth depends on the distance they have to span. To avoid any lateral movement in the joists of the upper floors, herring-bone strutting must be provided.

When the joists have been adequately fixed, the platform surface

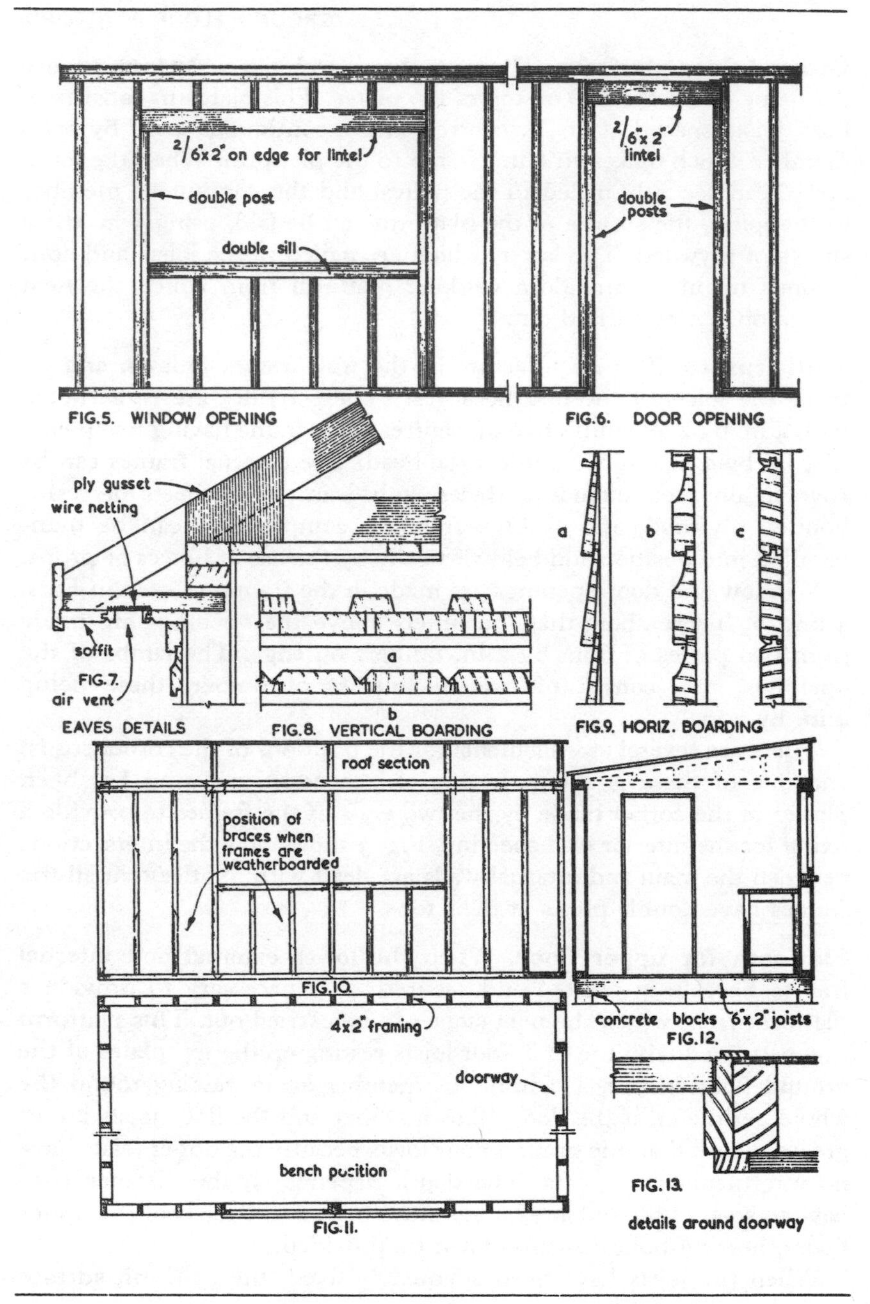
2/6"x2" on edge for lintel
double post
double sill
FIG.5. WINDOW OPENING
2/6"x2" lintel
double posts
FIG.6. DOOR OPENING
ply gusset
wire netting
soffit
FIG.7.
air vent
EAVES DETAILS
a
b
FIG.8. VERTICAL BOARDING
a
b
c
FIG.9. HORIZ. BOARDING
roof section
position of braces when frames are weatherboarded
FIG.10.
4"x2" framing
doorway
bench position
FIG.11.
concrete blocks
6"x2" joists
FIG.12.
FIG.13.
details around doorway

in the form of ½ in. plywood sheets can be fixed over the whole of the surface of the floor areas, as for the ground floor.

Upper floor frames. These are made on the platform, covered with the ½ in. thick sheets of resin-bonded plywood, and positioned and fixed to the other frames as they are completed.

Roof trusses. These can be constructed in the workshop and transported to the site when required. They are similar to those shown in Fig. 1, and can be constructed of 4 in. by 1½ in. timbers with ½ in. thick gusset plates glued or screwed together. These trusses should be spaced at 2 ft. centres, and when all have been positioned, ⅜ in. or ½ in. thick plywood sheathing is nailed to their top surfaces. A building paper which acts as a moisture barrier is laid over the whole plywood sheathing, and finally the roof finish is applied, in the form of cedar shingles or some other light material. If the traditional type of roof covering is contemplated it would be as well to select a roof truss designed to incorporate a purlin such as one of the TRADA roof trusses. Details of these can be found in the companion volume *(Practical Carpentry and Joinery)*, and in Chapter 6 of this volume.

A breather moisture barrier is applied to the sheathing on the outer wall frames, and over this is fixed the horizontal or vertical weatherboarding, as in Figs. 8 and 9. The outer walls of the house are insulated with fibre-glass quilt or similar material 2 in. thick fixed between the studs. A vapour barrier should be fixed on the edges of the studs facing the inside of the building and ½ in. plaster board placed over the barrier and fixed. The vapour barrier is necessary near the inside of the building to prevent moisture forming in the stud cavities.

Ceilings. Ground floor ceiling material can be fixed to the first floor joists in the normal way, but the ceilings to the upper floor will have to be fixed to 2 in. by 1 in. battens fixed at 16 in. centres across the lower edges of the roof trusses. A vapour barrier must be included between the ceiling material and the roof space, preferably on the top surface of the sheet material. To ensure adequate ventilation in the roof space, small ventilator slots should be made in the soffit of the eaves at short intervals round the building, see Fig. 7.

Site office. Carpenters are often asked to produce a site office or a shed which can be used as a store on site. Figs. 10, 11 and 12 show details of one which can be adapted for a store or site office. It consists of four frames, nailed together and covered on the outside with $\frac{5}{16}$ in. plywood sheets. If 4 ft. sheets are to be used the studs should be placed at 16 in. centres. If the outsides of the frames are not to be covered with ply, it will be advisable to include braces in the frames to keep them square. The outsides are generally covered with weather boarding. The 6 in. by 2 in. joists could be framed together in sections for easy transportation, and these could be laid on concrete blocks or sleepers which have been carefully levelled. The frames are positioned and bolted together at their corners. The roof could be made in one or more sections, laid on the tops of the frames and bolted in position.

9 Foot Bridges

From time to time the carpenter is called upon to provide a means of crossing an obstacle in the grounds of a factory or estate to enable people to go from one point to another with the minimum of delay and to prevent the risk of accident. Perhaps the commonest obstacles in the grounds of a factory are pipes, Fig. 3, and if one of these is in such a position as to impede the regular movement of workmen, a safe means of crossing the obstacle should be provided. This would necessitate the manufacture of a small footbridge. That shown in Fig. 3 consists of four posts into which rails are tenoned to support the platform immediately above the pipe, and on each side are a set of steps consisting of strings and treads. The strings are housed to take the ends of the treads, and these are secured by three bolts passing through the steps and connecting the two strings. Rails can be used for stopping the spread of the lower ends of the steps.

Newels and safety rails should be provided on one side if the platform is only a short distance from the ground, and on both sides if its height is, say, over 4 ft. Cross braces should be used between the posts in the direction of the length of the pipe, as seen in Fig. 4, to keep the whole job rigid and square.

Footbridge over stream. Where a stream or some sort of excavation has to be safely crossed, a footbridge somewhat on the lines of that shown in Figs. 1 and 2 could be used. Two beams should be constructed, incorporating the posts supporting the safety rails. Two timbers are used for both top and lower chord, and ties, positioned as in Fig. 1, are included so that one bolt can be used at each intersection for securing the timbers together. The only points where this cannot be done are at the centre. Packing pieces will have to be used at these points.

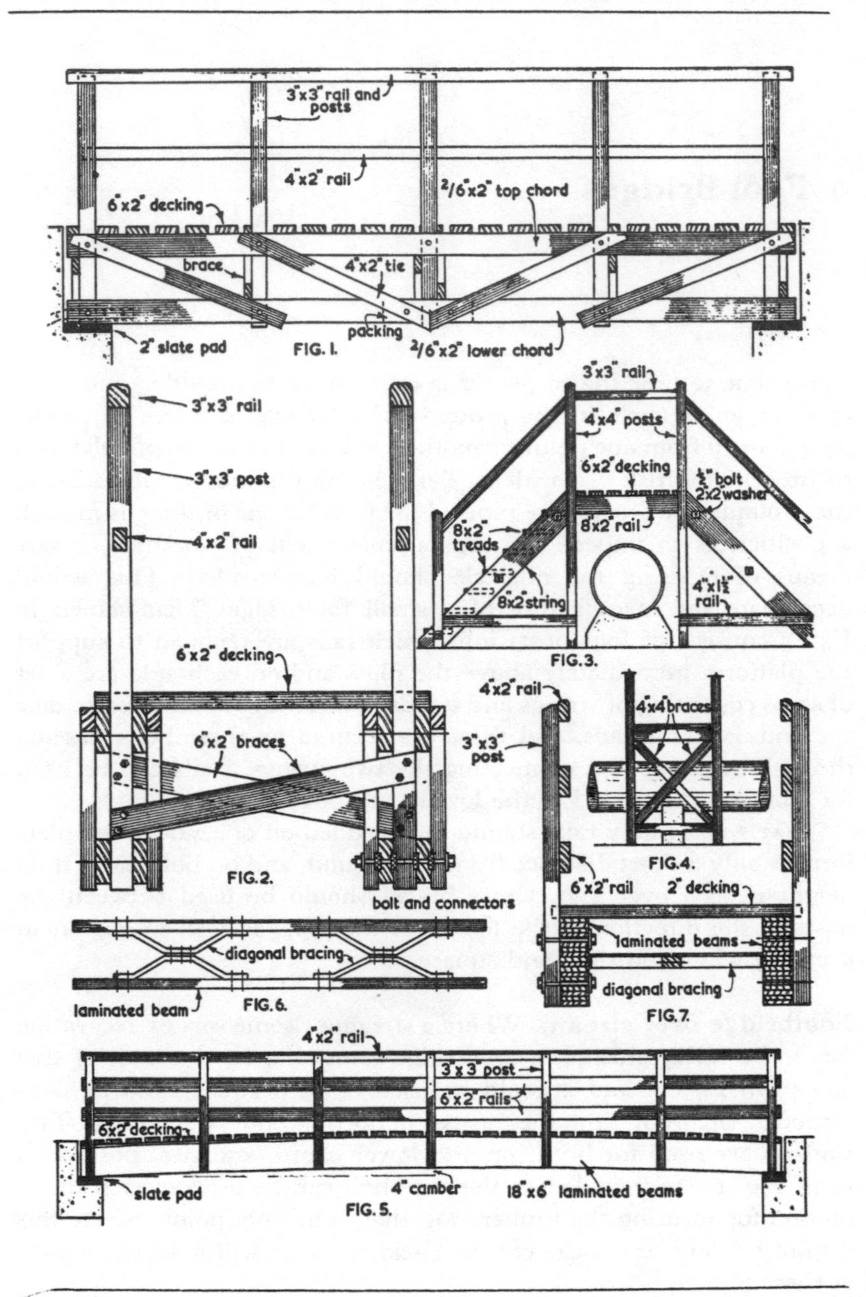
3"x3" rail and posts
4"x2" rail
2/6"x2" top chord
6"x2" decking
brace
4"x2" tie
packing
2" slate pad
FIG. I.
2/6"x2" lower chord
3"x3" rail
3"x3" post
4"x2" rail
6"x2" decking
6"x2" braces
FIG.2.
bolt and connectors
3"x3" rail
4"x4" posts
6"x2" decking
½" bolt
2x2 washer
8"x2" treads
8"x2" rail
8x2 string
4"x1½" rail
FIG.3.
4"x2" rail
4"x4 braces
3"x3" post
FIG.4.
6"x2" rail
2" decking
laminated beams
diagonal bracing
FIG.7.
diagonal bracing
laminated beam
FIG.6.
4"x2" rail
3"x3" post
6"x2" rails
6x2 decking
slate pad
4" camber
18"x6" laminated beams
FIG. 5.

Fig. 2 is a cross-section through the footbridge and illustrates how the beams are made up. It also shows that the beams are cross-braced to minimise movement of the bridge when in use. The drawing also shows that each joint is secured with a bolt with a timber connector between each pair of timbers. Large square washers should be used beneath the bolt heads and nuts.

Beams support the decking, which should be fixed with a space between each board. Top safety rails should be 3 ft. 6 in. from the top of the decking. Intermediate safety rails should also be used and can consist of one or more pieces equally spaced between the top rail and decking.

Base. The bridge should stand on properly prepared bases, the timber in contact with slate pads which are bedded on to concrete. Mild steel straps can be included in the latter when this is being poured to secure the bridge in position.

Longer footbridge. Figs. 5, 6 and 7 show details of a more elaborate type of footbridge often found in public parks, etc. The beams in this case are made up from laminated timbers, and have a small camber in their lengths. These beams are made in a similar way to those shown in Chapter 18. The construction of this type of bridge is much the same as that shown in Fig. 1. The posts supporting the safety rails are bolted on the sides of the beams, and the bridge is supported and secured as before. Cross-bracing has to be provided, as shown in Fig. 6. It consists of angle iron welded to the shapes shown and bolted to the inside surfaces of the beams. Each half is bolted to the other at their centres, and with a hardwood packing at the joint.

10 Doors

Work involving doors will always play a big part in the life of the carpenter and joiner. The common types of door were dealt with in the companion volume *(Practical Carpentry and Joinery)* but there remain several which must be included in a book on advanced work.

Double-margin door. Where a single door has to be fitted to a rather wide opening, a double-margin door is often used, Fig. 1. In appearance it looks like a pair of doors. On the right of the drawing is shown the method used for constructing and assembling it. As can be seen, four stiles are used, the middle two being secured together by means of pairs of folding wedges.

The two halves should first be constructed as though they were to be used as double doors. Mortices are made across the width of the meeting stiles in line with each other. The doors are first assembled without their panels as separate units, glue being used on the mortices and tenons to the centre stiles only. The two halves are brought together and secured with the folding wedges through the centre stiles. When the ends of the wedges have been trimmed, the outside stiles are removed, the panels inserted into their grooves, the outer stiles glued, and the whole reassembled. The drawing on the right shows the panels being assembled into the nearly completed door. Fig. 2 is a cross section through the meeting stiles.

Vestibule screen. Fig. 5 shows a vestibule screen with fixed side lights and double swing doors. These are situated just inside the main entrance to a building, such as a large store, show rooms, hotels, etc. The screen shuts off the main part of the building from the outside. The elevation shows that the screen is fully glazed and has a transom.

Fig. 6 is a horizontal section through the screen, and Figs. 7 and 8

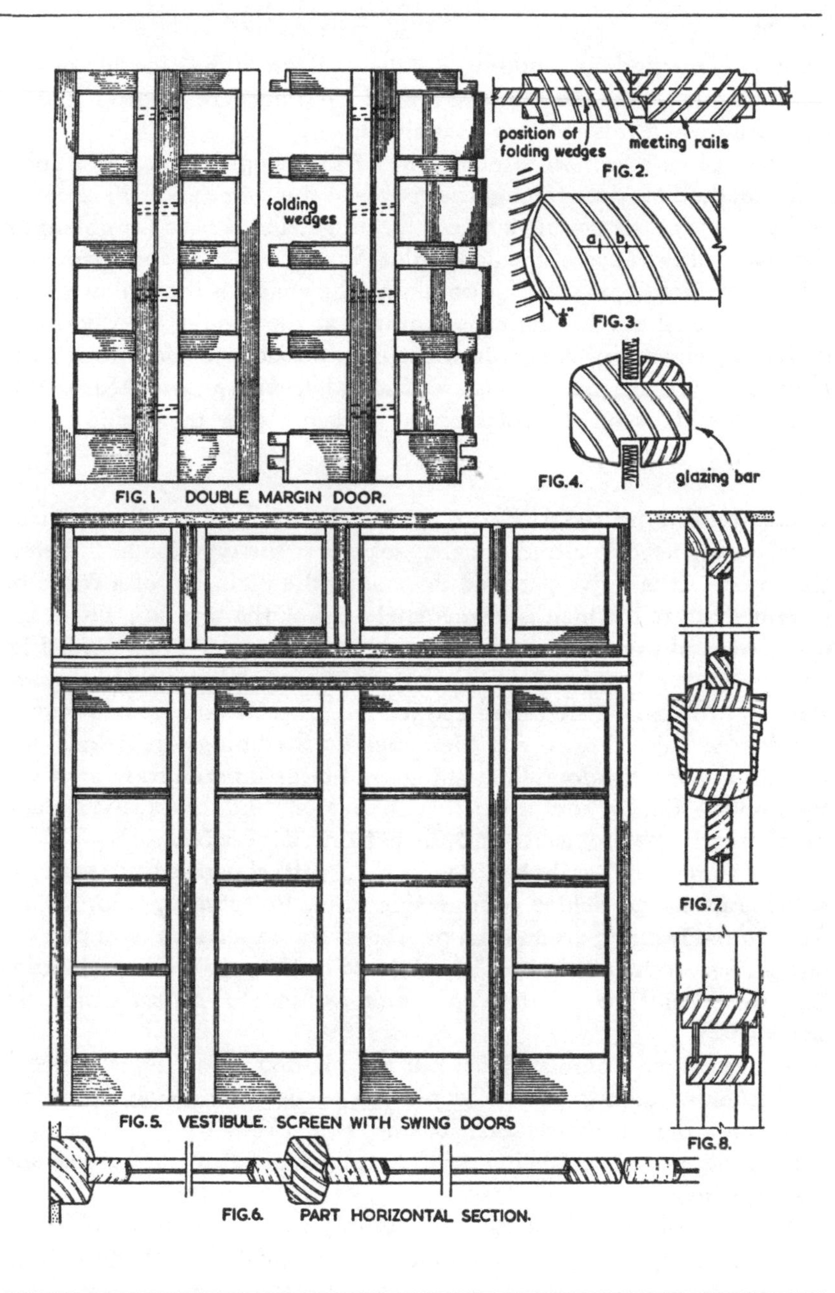

FIG. 1. DOUBLE MARGIN DOOR.

FIG. 5. VESTIBULE. SCREEN WITH SWING DOORS

FIG. 6. PART HORIZONTAL SECTION.

show two methods of building up the transom. The side lights and the fanlights above the transom are all fixed and are incorporated in the frame when this is being assembled.

Fig. 3 gives a section through one of the hanging stiles of a door, and a method for determining the curves of the frame and stile. Point b is the centre of the double-action floor hinge, and is used for obtaining the shape of the edge of the door stile. Point a, which is approximately ½ in. away from b, is used for obtaining the shape of the hollow in the door frame. About ⅛ in. joint is required at the edge of the door, and if the curves are set out as shown it is impossible to see through the joint. Fig. 4 is a section through a glazing bar with glazing beads. The beads should be on the sides of the screen facing the inside of the building.

Traditional type door. Fig. 9 is the elevation of another internal door, constructed in a traditional style, and is the type found in better class work. It is a five-panelled door with the inclusion of a frieze or intermediate rail, which is immediately below the top rail. There are many ways of preparing the framework and panels of a door, and Fig. 10 shows several of these. At a the door framework is moulded on each side, and the panels are raised and fielded on each side. The linings of the door are also framed and the panels to the linings are fielded and raised to match the door. The linings and built-up architrave are fixed to grounds. The architrave extends down to a plinth block, Figs. 9 and 11, to which the architrave and the skirting are scribed.

The framework at b has its framing left square-edged with its panels raised and fielded on one side only. Bolection mouldings are fixed to the framing around the panels on the raised side, and planted mouldings on the rear side. The methods used for fixing these mouldings were described in the companion volume, *Practical Carpentry and Joinery*.

At c is shown a similar finish but the planted moulding is replaced by a stuck moulding. Screws would not be used for fixing the bolection moulding in a case such as this, oval brads or lost-head nails being used, their heads being carefully punched below the surface and the holes filled.

At d are details of a door with double-raised and fielded panels, and framed double bolection mouldings. The bolection mouldings and panels would be made up as single units, mitred, tongued, and

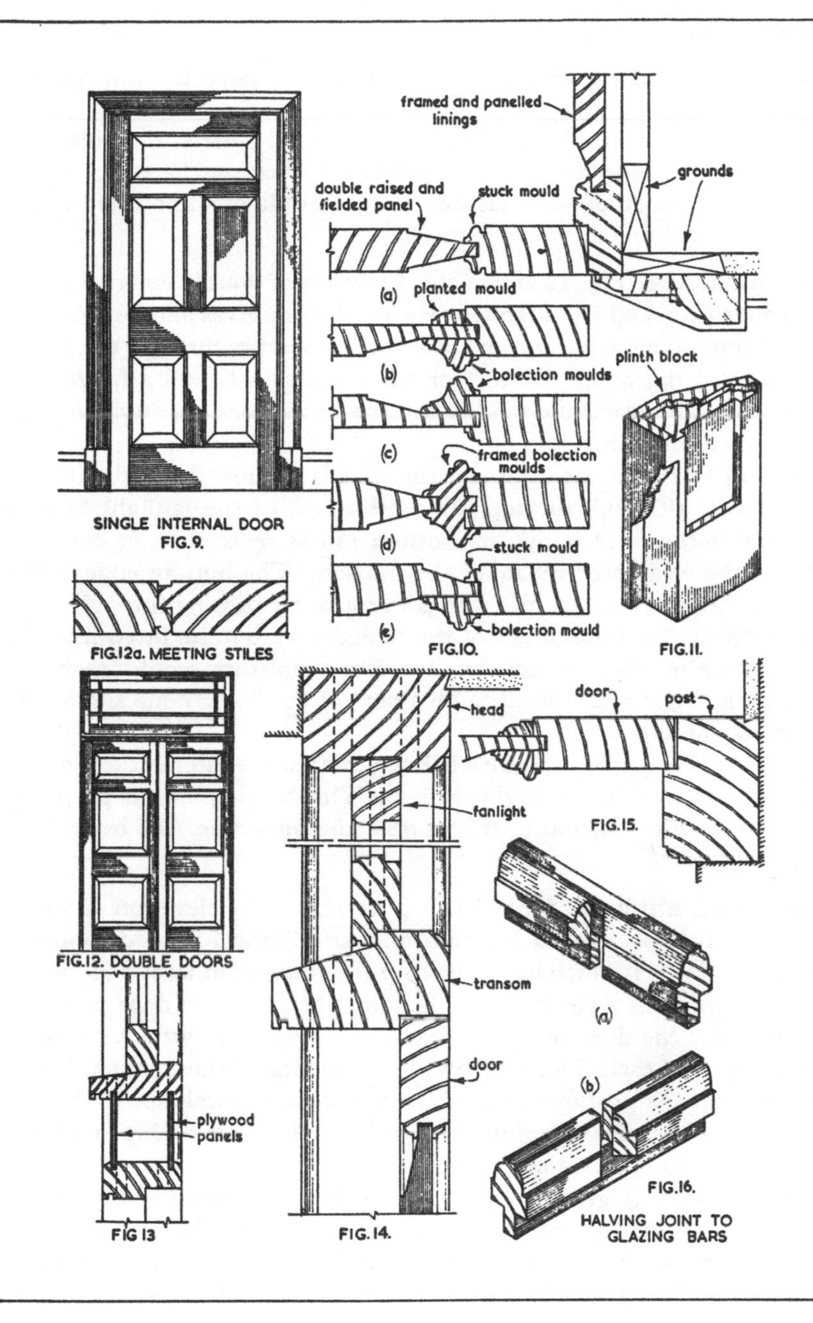
framed and panelled linings
grounds
double raised and fielded panel
stuck mould
(a)
planted mould
(b)
bolection moulds
plinth block
(c)
framed bolection moulds
(d)
stuck mould
(e)
bolection mould
SINGLE INTERNAL DOOR
FIG. 9.
FIG. 12a. MEETING STILES
FIG. 10.
FIG. 11.
head
door
post
fanlight
FIG. 15.
(a)
FIG. 12. DOUBLE DOORS
transom
door
(b)
plywood panels
FIG. 16.
HALVING JOINT TO GLAZING BARS
FIG 13
FIG. 14.

screwed at their corners, and included in the door framing during assembly.

At e the door has double-fielded and raised panels, the sides not matching. One side has stuck mouldings worked on the edges of the frame, and the other side has bolection mouldings. There are many variations to those shown in Fig. 10.

External doors. Fig. 12 shows the elevation of double external doors with a transom and fixed fanlight above. Fig. 13 gives another method of building up a deep transom, and 12a is a section through the meeting stiles of the doors. Each door has three panels and a frieze rail. The panels are raised and fielded and the framework has stuck moulding worked on each side.

Fig. 14 is an enlarged view of the details around the fanlight, and shows that, although the top rail (and stiles) of the fanlight fit into recesses in the door frame, its bottom rail is rebated to fit over the edge of the weathered rebate of the transom. The bottom edge of the rail has a groove worked into it to prevent water passing through to its back edge. If the bottom rail of the fanlight were to be fitted into the top surface of the transom, trouble from moisture would probably result. Fig. 15 is a section through a post of the door frame and a stile of one of the doors.

Often it is required to make a halving joint at the intersections of the glazing bars, as in the fanlight, Fig. 12. This halving joint is prepared as shown in Fig. 16a and b. If required, the mouldings can be scribed instead of mitred.

Doors with elliptical fanlight. Fig. 17 shows the elevation of a pair of external doors, with a transom and an elliptical-headed fanlight. In such a case it is much better to keep the transom at least 2 in. below the springing line so as to allow the joints between the door post and transom and the door post and head to be prepared without affecting the strength of each. Fig. 18 shows the positions of these joints. It can be seen that the post of the door frame extends to well above the top edge of the transom, keeping the two joints well apart. Fig. 19 gives the types of joints used at this point.

Figs. 20 and 21 give two ways of finishing the middle rails of the doors. The first, Fig. 20, has square shoulders to the rail which is made up in two halves so that the circular panel can be assembled. In the second method, Fig. 21, the rail has splayed shoulders as well as

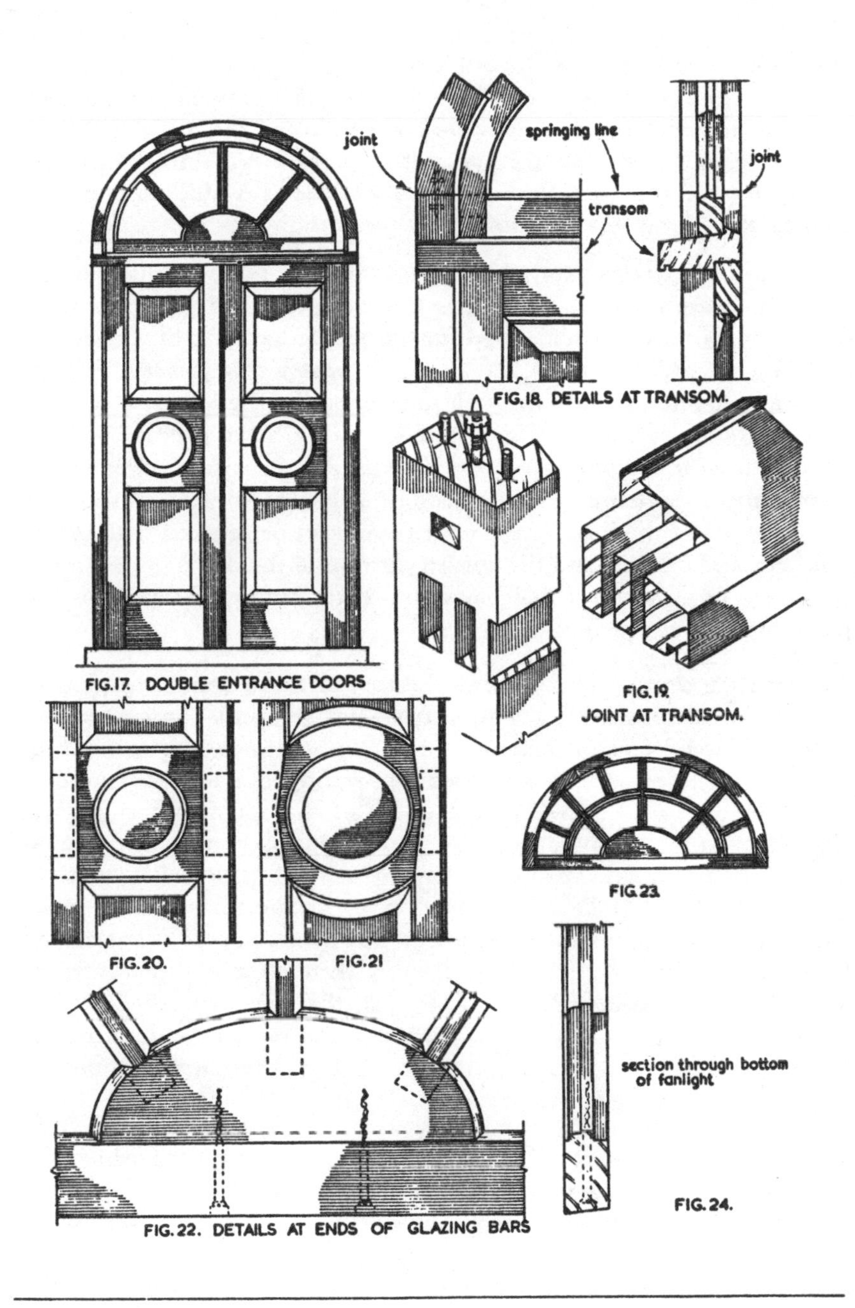

FIG. 17. DOUBLE ENTRANCE DOORS

FIG. 18. DETAILS AT TRANSOM.

FIG. 19. JOINT AT TRANSOM.

FIG. 20.

FIG. 21

FIG. 22. DETAILS AT ENDS OF GLAZING BARS

FIG. 23.

FIG. 24.

curved top and bottom edges, allowing for a larger circular panel.

Fig. 22 shows the bottom of the fanlight. To prevent complicated joints at this point, and to improve on the look of the work, a shaped piece is fixed to the top surface of the bottom rail, into which the lower ends of the bars are morticed. Fig. 24 is a side view of the shaped piece, and Fig. 23 shows another fanlight design.

Pointed-arch doors. Church doorways usually follow a shape based on pointed-arch design. Fig. 25 is the elevation of a pair of church doors based on the methods used for framed, ledged and braced doors, see the companion volume, *Practical Carpentry and Joinery*. Fig. 26 gives a horizontal section through the doors. To the left is an outside elevation and on the right is an inside view showing cross bracing which assists in keeping the doors square. All the inside edges of the frames are stop-chamfered, and each door is cross-braced as shown.

The joints of the head of the door frame can be secured with handrail bolts and dowels, and the curved portion of the doors joined at the springing line by using the joints shown in Fig. 27. At the top the usual mortice and tenon joint is used.

Tudor arch door. Fig. 29 shows the top part of a door which has to fit into an opening with a Tudor arch. These joints can be secured with handrail bolts or hammer-head key joints as shown in Fig. 30a and b. The method employed for setting out a Tudor arch is shown in Fig. 28. Let b–d be the width of the opening, and a–c the rise. Draw the rectangle a–3–b–c and divide b–3 into three equal parts. Join 2 to a, and with centre b and radius b–2 describe an arc to give 2′ on b–c. Make a–2″ equal b–2′ and bisect 2′–2″. Make 2–a–x a right angle, and extend a–x downwards until it meets the bisecting line in x. Draw a line from x through 2′. With centre 2′ and radius 2′–b describe the first part of the curve from b round to the line x–2′ to give point z. With centre x and radius x–a describe the other part of the curve a–z. The other half of the curve is found in exactly the same way as the first half.

Flame-resisting doors. Occasionally doors are required which are capable of resisting flames for a period of at least 30 minutes. There are several ways of making them, and two are seen in Figs. 31 and 33. Fig. 31 is the elevation of a panelled door, and the section through the door, Fig. 32, indicates that the panels are the same thickness as the

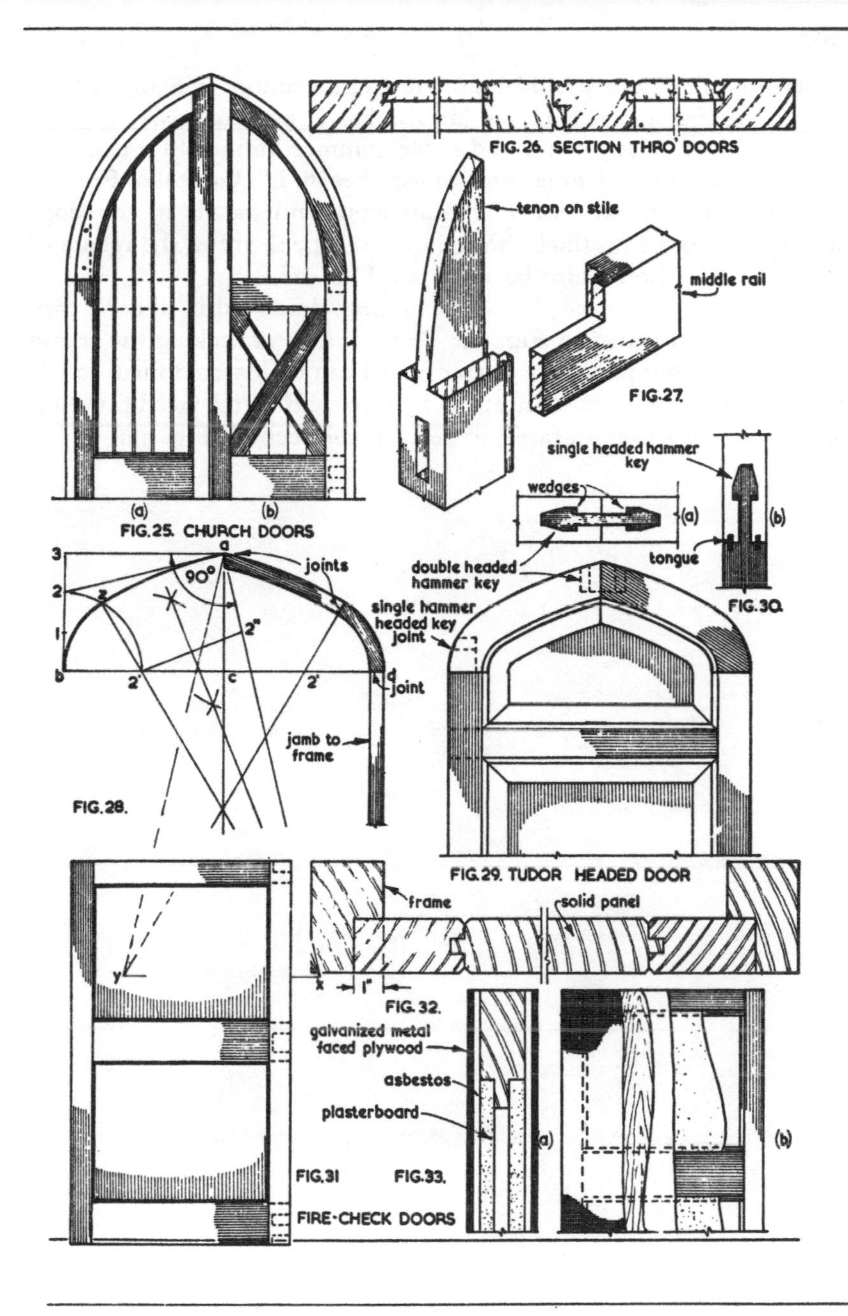
FIG. 26. SECTION THRO' DOORS
tenon on stile
middle rail
FIG.27.
(a)
(b)
FIG.25. CHURCH DOORS
single headed hammer key
wedges
(a)
(b)
tongue
FIG.30.
3
2
1
b
a
90°
joints
2"
2'
c
2'
d
joint
jamb to frame
FIG.28.
double headed hammer key
single hammer headed key joint
FIG.29. TUDOR HEADED DOOR
frame
solid panel
y
x
1"
FIG. 32.
galvanized metal faced plywood
asbestos
plasterboard
(a)
(b)
FIG.31
FIG.33.
FIRE-CHECK DOORS

framework. The door should be made from a timber which has high natural fire resistance such as jarrah, karri, padauk, teak, gurjun, and greenheart. Many other and more common timbers are resistant to fire but to a lesser degree, such as ash, beech, idigbo, iroko, English oak, sycamore, and so on. The framework and panels of the door should be at least 2 in. thick throughout and the depth of the rebate in the door frame should also be 1 in., see Fig. 32.

Another method used for constructing doors with a good fire resistance is illustrated in Fig. 33. This consists of making the frame of the door so that panels of plaster board can be inserted flush to the two surfaces of the frame. Sheets of asbestos and metal-faced plywood are then secured to both faces, as seen in the section, Fig. 33.

11 Sliding Door Gear

There are many advantages in using modern sliding door gear. Apart from the ease in opening and closing, especially with large doors to garages, warehouses, etc., and the saving of space within the building, almost absolute silence is obtained. The drawings in this chapter are reproduced by permission of P. C. Henderson Ltd., of Romford, Essex, who are the manufacturers of gear for almost every form of doorway.

Room doors. Figs. 1 to 6 show details of the *Marathon* ball-race gear suitable for doors in houses and flats up to a weight of 150 lb. It is designed for internal doors but can, if required, be used externally if properly protected. The sets as purchased from the manufacturers consist of the complete top assembly, a bottom channel for fixing to the bottom surface of the door, a floor guide, and one end or centre stop. If special fittings are required the manufacturers will supply them.

Fig. 1 illustrates the common method used for fixing the *Marathon* gear. The top assembly is screwed to a board which is fixed above the door opening, and a metal pelmet is fixed above the gear to hide it from view. At the bottom, the lower edge of the door is grooved to receive the metal channel which is secured by screws into the woodwork. The floor or bottom guide is of nylon and is screwed to the floor and in conjunction with the channel keeps the door in correct alignment. There should be a $\frac{3}{16}$ in. joint between the floor and the bottom of the door.

Fig. 2 shows the gear fixed directly to the lintel above the doorway, and a wooden pelmet is used in place of the metal one, which is supplied by the manufacturers as an optional extra. Fig. 3 shows the type of bracket required when the gear is to be fitted to the soffit of the opening. If double doors are required in this position, the brackets are fixed side by side as shown in Fig. 5. Fig. 4a is a view of the bottom nylon guide, which, is fixed near the jamb of the opening. Fig. 4b and c show alternative bottom guides to the door when these are favoured. Fig. 6 is a front view of the track and ball race.

Roller gear. Fig. 7 is a cross-section through the *Majestic* bottom roller gear. This is ideal for high class joinery items, and is most suitable for picture windows, Fig. 12, and doors to showrooms, etc. It is used for doors with a minimum thickness of 1⅞ in. and a maximum weight of 600 lb. per door. The brackets holding the top box guide through which the rollers pass are fixed to the face of the masonry as shown, and if double doors are used, the brackets shown in Fig. 8 must be obtained. When double doors are to be placed beneath the soffit of the opening, the two box guides, Fig. 9, can be fixed to the woodwork covering the soffit.

The bottom rollers, which take the weight of the door, pass over a brass rail which can be recessed into a hard-wood or concrete sill, see Fig. 7. An alternative to this arrangement is seen in Fig. 10, where a brass rail, also used as a weather bar, has been let into the top surface of the weathered sill. To prevent draughts through the sides of the doors, sponge rubber buffers can be used, as in Fig. 11.

Small doors. For cupboard, cabinet and wardrobe doors up to a weight of 50 lb. per door *Loretto* cabinet rollers can be used, Figs. 14 and 16. These are so made that condensation will not affect them, and they are smooth and silent in their action. Fig. 14a, b and c show top guides with retractable shoots or bolts which enable the doors to be removed with the minimum of trouble. Fig. 16 gives two types of nylon top guides which are not retractable. That on the left is for edge fixing and that on the right is fixed from the top.

For frameless glass doors to cabinets, showcases, etc. an economical method of sliding the doors is by means of the track, Fig. 15d, and slider, Fig. 15a.

Garage doors. For two-leaf garage doors up to a maximum weight of 260 lb. per door, and where the doors are required to slide to each side of the opening as in Fig. 19, the straight sliding gear seen in Figs. 17 and 18 can be used. This consists of fixing brackets, which hold the track, to the inside or lower surface of the lintel above the

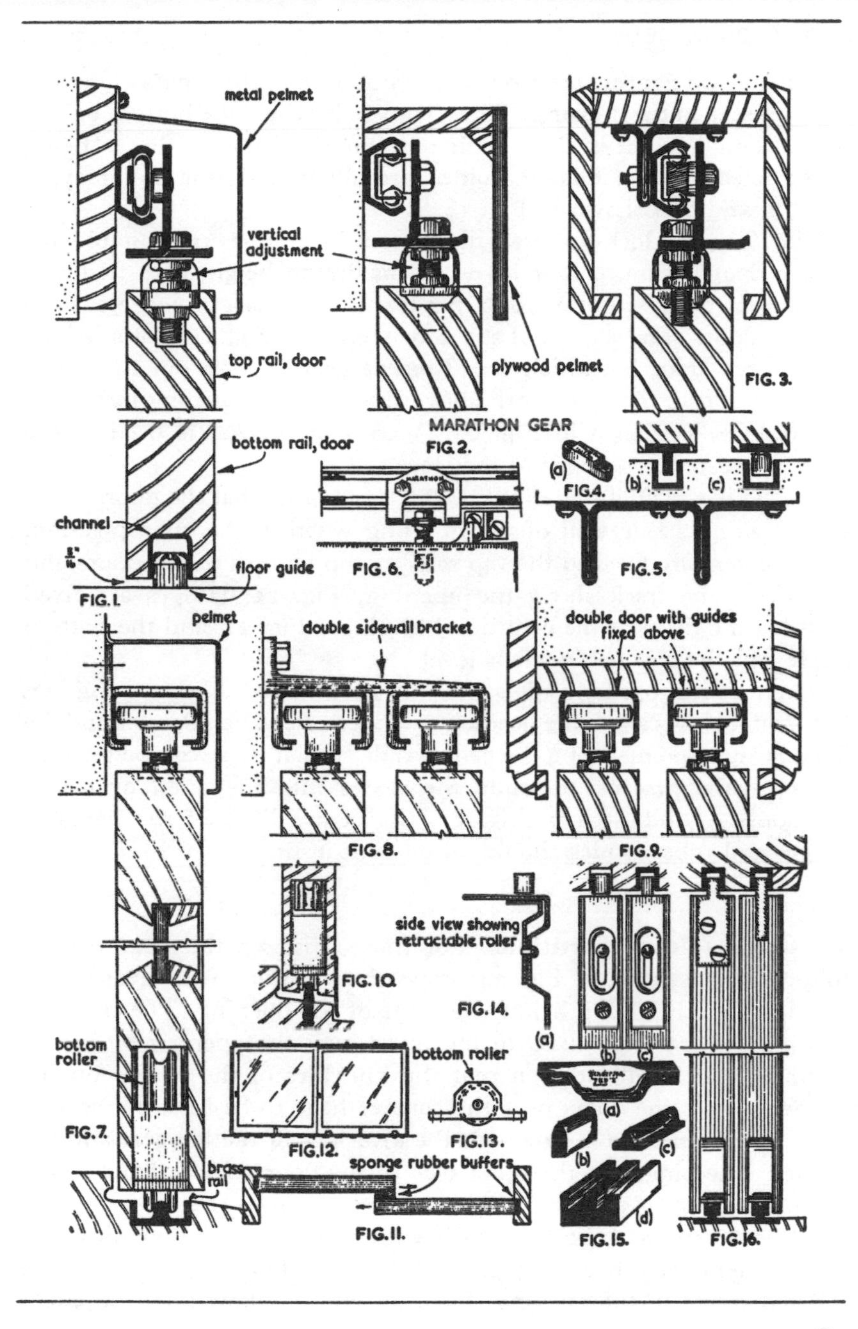

metal pelmet
vertical adjustment
top rail, door
plywood pelmet
FIG. 3.
MARATHON GEAR
FIG. 2.
bottom rail, door
channel
floor guide
FIG. 1.
(a)
FIG. 4.
(b)
(c)
FIG. 6.
FIG. 5.
pelmet
double sidewall bracket
double door with guides fixed above
FIG. 8.
FIG. 9.
side view showing retractable roller
FIG. 10.
FIG. 14.
(a)
(b)
(c)
bottom roller
FIG. 7.
bottom roller
FIG. 13.
FIG. 12.
sponge rubber buffers
brass rail
FIG. 11.
(d)
FIG. 15.
FIG. 16.

opening. Two hangers are fixed to each door, and these pass along the track, as seen in the drawings. The channels in which the bottom guides travel should be let in flush with the concrete floor as this is being constructed. Two kinds of guides are illustrated, the roller type, Fig. 17, and that shown in Fig. 18.

The *Tangent* lock-up gear illustrated in Fig. 21 is suitable for garage doors of up to four leaves, a maximum height of 7 ft. 6 in., and a maximum weight of 70 lb. per leaf. Fig. 20 is a typical elevation of such doors. The section of the door shown in Fig. 21 is of a ledged and braced door for which the *Tangent* gear is most suitable. The track is fixed to the woodwork above the doorway opening by means of brackets, as shown in Fig. 22. A coupler bracket is used at the joint between two sections of the track.

Fig. 23 is a plan of the arrangement, and shows that the doors come to rest along a side wall of the building when in the open position. The hangers are fixed to the top rails or top battens of the doors and run along the track above the opening, Fig. 22. Hinges are fixed between the doors at the centre rail level, see Fig. 21, and the bottom guide and the channels at floor level.

The first leaf opens in the same way as a hinged door, see Fig. 23, this leaf being called the leading swinger. Bow handles should be fixed at three points along the centre rails so that these can be gripped to apply the pressure to slide the doors into the open or closed positions. As indicated in Fig. 21, the centre lines of the hangers, hinges and roller guides should all be kept in line.

Top-hung folding partitions. For these, Fig. 25, the *Council* end-folding gear is suitable. Fig. 24 shows the position of the gear where the doors are not hung below the soffit of the opening. The brackets holding the track are fixed to the lintel over the opening by means of bolts, and it will be seen that the knuckles of the hinges on the hanger side of the doors near the centres have to be kept out so that their centre lines are in line with the knuckles of the hangers and the guides. The hinges on the other face of the doors can be fixed in the normal fashion.

If the doors are to be kept flush with one face of the opening, the arrangement seen in Fig. 27 can be adopted. This involves having a rebated frame in the opening. Figs. 28 and 30 show how the edges

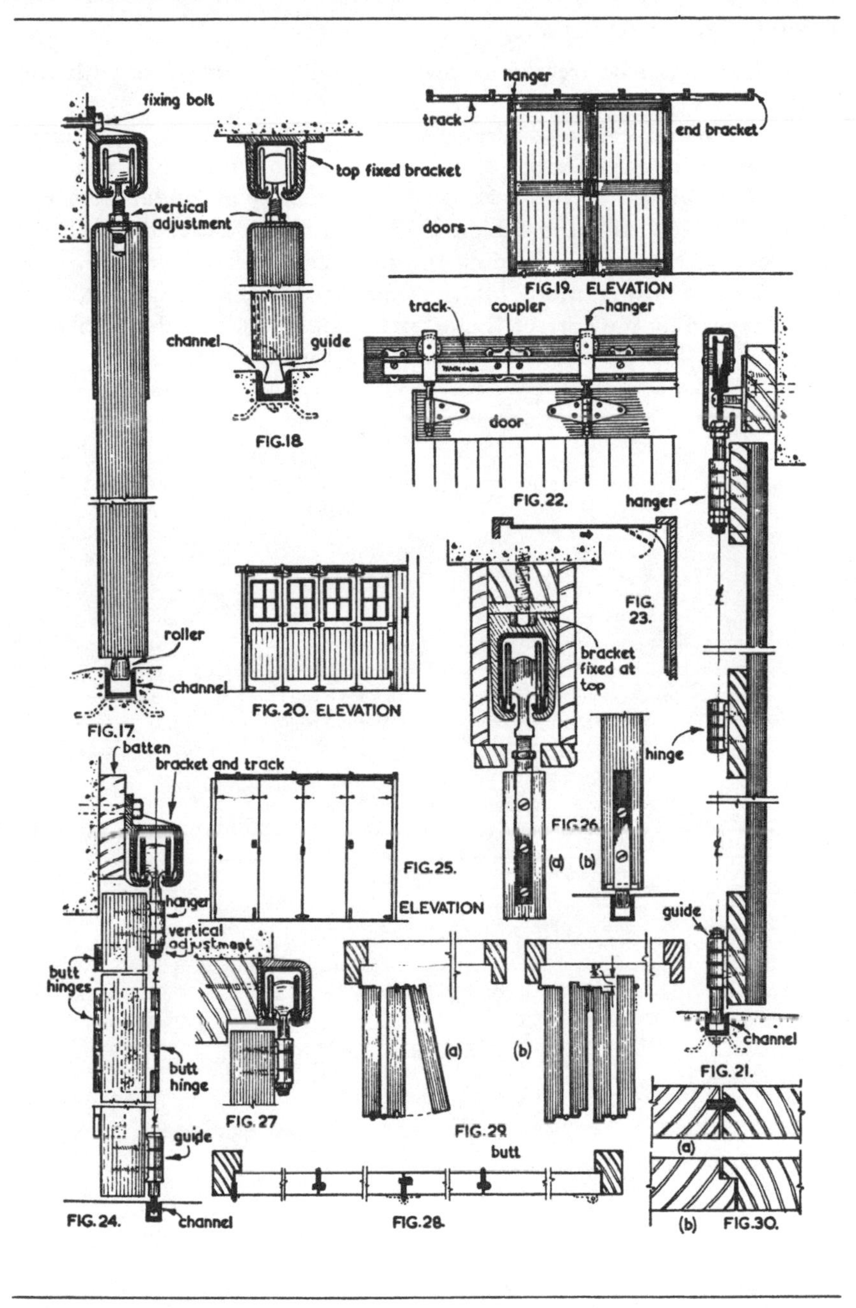
fixing bolt
vertical adjustment
roller
channel
FIG.17.
top fixed bracket
vertical adjustment
channel
guide
FIG.18
hanger
track
end bracket
doors
FIG.19. ELEVATION
track
coupler
hanger
door
FIG.22.
FIG.20. ELEVATION
FIG. 23.
bracket fixed at top
hanger
hinge
guide
channel
FIG. 21.
batten
bracket and track
hanger
vertical adjustment
butt hinges
butt hinge
guide
channel
FIG. 24.
FIG.25. ELEVATION
FIG.26
(a) (b)
FIG. 27
(a)
(b)
FIG.29
butt
FIG.28.
(a)
(b) FIG.30.

of the leaves can be treated so that they will not interfere with the hinges.

When the end leaf is a swinger, similar to a normally-hinged door, all the leaves of the partition should be equal in width, Fig. 29a, but where the end leaf carries a hanger it must be $\frac{3}{4}$ in. wider than the others, Fig. 29b. If the partition is to be directly under the soffit of the opening the brackets holding the track can be fixed as in Fig. 26.

Readers requiring more information on specific door gear, and advice regarding the correct fixing, etc., should write to the manufacturers, P. C. Henderson Ltd., Romford, Essex.

12 Windows with Curved Heads

Figs. 1, 2 and 3 show details of a semi-circular-headed frame with vertical sliding sashes. On the left of Fig. 1 the outside linings have been removed to show how the pulley stile has been fixed to the built-up head. The latter, which is semicircular, has been built up in three thicknesses, see Fig. 2, the centre section being the continuation of the parting bead, which is approximately $\frac{3}{8}$ in. thick and projects $\frac{5}{8}$ in. beyond the surfaces of the other two sections. The various segments of the head are glued and screwed together with the joints staggered.

The pulley stile is allowed to run up to well above the springing line of the frame, and is glued and screwed to the prepared flat surface of the head. A block shaped to fit between the stile and the head can, if necessary, be glued and pinned in position, as shown in Fig. 1. The two outer sections of the head are stopped about 4 in. above the springing line as the ends are required to act as stops for the sash. A small horizontal surface to correspond to the stop is provided, as in Fig. 7.

Fig. 2 is a section through the top portion of the frame and sashes, and it can be seen that blocks have been cut to fit between the linings above the built-up head to strengthen the work at this level. Fig. 3 is a horizontal section through the frame and a sash. Details around the sill are as shown in the companion volume, *Practical Carpentry and Joinery*. If it is required to have sliding sashes which act as pivot-hung sashes in addition to sliding vertically, double stiles must be provided for the sashes with stops between. These are then prepared as for the pivot-hung sash stops described in the companion volume.

Semicircular-headed frame with transom. Fig. 5 is a pictorial view of the top portion of a semicircular-headed frame involving a

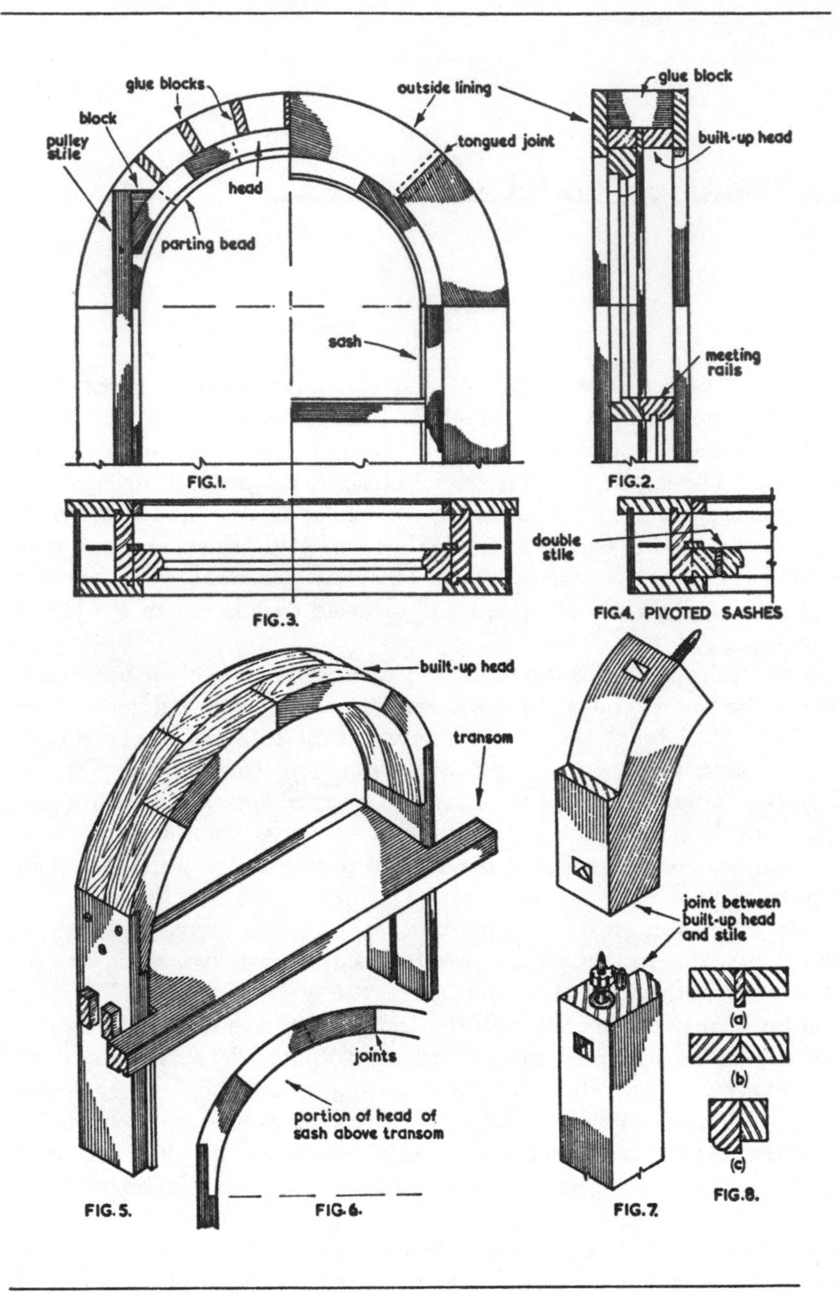
glue blocks
block
pulley stile
head
parting bead
outside lining
tongued joint
sash
glue block
built-up head
meeting rails
FIG.1.
FIG.2.
double stile
FIG.3.
FIG.4. PIVOTED SASHES
built-up head
transom
joints
portion of head of sash above transom
joint between built-up head and stile
(a)
(b)
(c)
FIG.5.
FIG.6.
FIG.7.
FIG.8.

transom and with sliding sashes below. The head of the frame in this case is built up of two sections only; a parting bead not being required above the transom. The latter is weathered on its top surface, and is kept down below the springing line so as not to complicate the joints around that section of the frame. Fig. 6 shows how the top part of the sash above the transom would be made.

Fig. 8 shows how the various shapes of heads of curved frames and sashes can be built up. At the top, a, are the three sections making up the head of the window in Fig. 1; b shows how the head of the frame in Fig. 5 is built up; and c indicates how the rebate can be formed in the semicircular portion of the sash shown in Fig. 6.

13 Roof Lights and Ventilators

Roof lights can be in the form of skylights, dormer windows, eyebrow windows and lantern lights.

Skylight. This consists of an opening in a roof, usually small, with a curb inserted around the opening. On top of the curb is placed the glazed sash which allows light to pass into the space below, see Fig. 1. The opening or trimming in the roof is made in the same way as an opening for a chimney, two of the common rafters and the trimming pieces forming the rectangular opening into which the curb is fitted.

Curb. The curb should be prepared from, say, 1¼ in. or 1½ in. material, and can be either butt-jointed at the corners, or, if it has to be moulded on the lower inside edges, mitred or scribed at these points. The curb should be allowed to extend upwards so that its top edges are at least 3 in. above the surfaces of the roof covering, enabling it to be weathered in the correct manner, as seen in Figs. 1 and 2. The light is made so that it projects well beyond the four edges of the curb.

Condensation will inevitably settle on the lower surfaces of the glass panes at one time or another, and provision has to be made to allow this moisture to escape on to the roof surface. The top surface of the bottom rail of the light is recessed so that the panes of glass pass over and rest on only a portion of the top surface. Condensation grooves or recesses are made for each pane, approximately ⅛ in. deep, allowing the moisture to escape between the glass and the bottom rail, Fig. 3.

The light can be made to open, which means that it will be hinged at its highest point, or it can be fixed. In the latter case it can be screwed down on to the curb from the top and the screw holes either pelleted or filled with putty. A close look at the drawings shows the

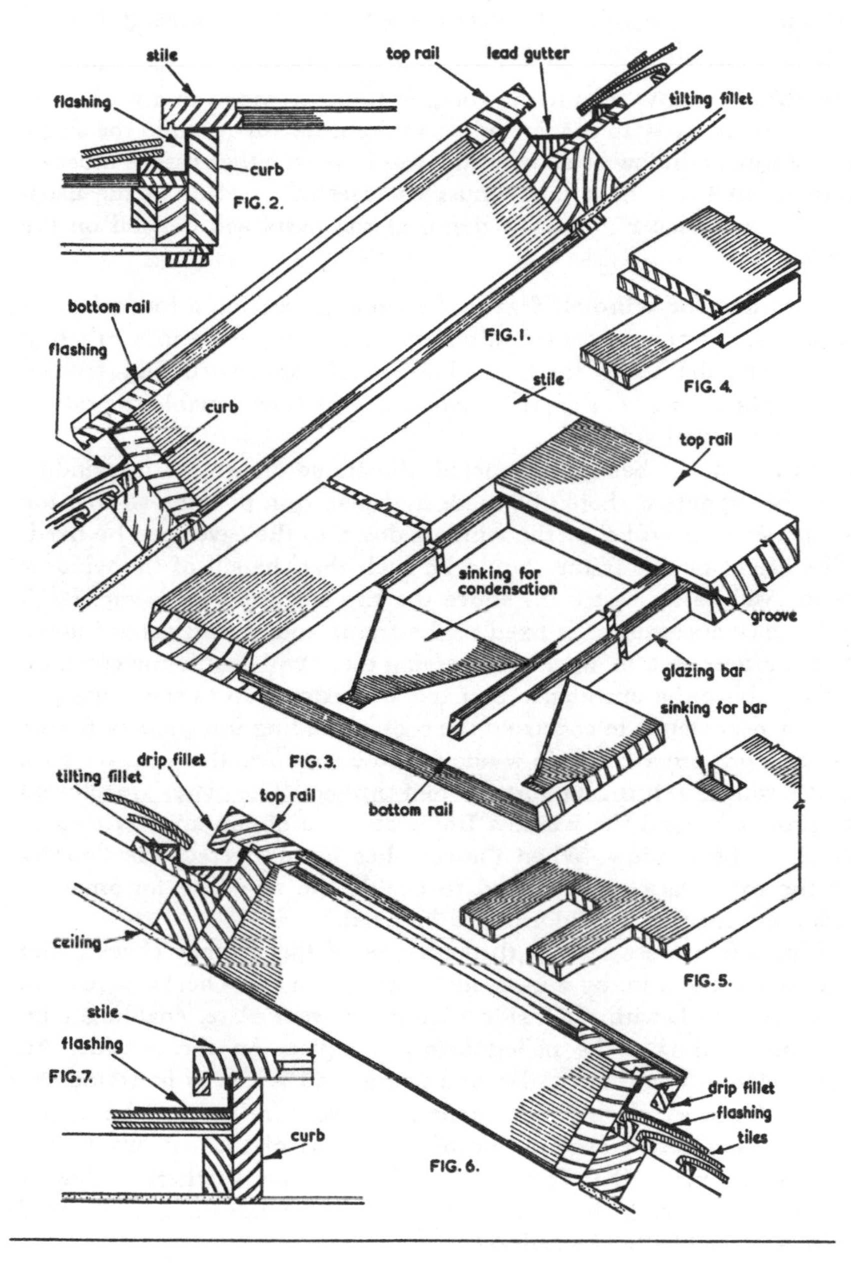
stile
flashing
curb
FIG. 2.
top rail
lead gutter
tilting fillet
bottom rail
flashing
curb
FIG. 1.
FIG. 4.
stile
top rail
sinking for condensation
groove
glazing bar
sinking for bar
drip fillet
tilting fillet
top rail
FIG. 3.
bottom rail
ceiling
FIG. 5.
stile
flashing
FIG. 7.
curb
FIG. 6.
drip fillet
flashing
tiles

additional work involved to ensure a watertight job. Figs. 3–5 show how the skylight is constructed.

Dormer window. If a living room is to be provided in a roof space it is also necessary to provide adequate light for the room in the shape of a dormer window or an eyebrow window. In either case it must be remembered that floor joists must be inserted in place of the usual ceiling joists, see Fig. 7. The depth of the joists will depend on the span.

Allowance for window frame. When constructing a roof which is to include a room, the roof rafters must be supported in a different way from the usual method. This involves constructing trussed or boxed purlins, see Fig. 16, Chapter 6, which will enable the rafters to be supported vertically.

As the roof is being constructed, the frame on which the window is to be supported should be made and placed in position so that the short rafters from below the window down to the eaves can be fixed. The height of this frame should be such that the sill of the window frame will be at least 6 in. above the top surfaces of the rafters. A ¾ in. thick apron piece is fixed to the frame and should extend down to the rafters. A 4 in. by 3 in. trimming piece will enable the common rafters above the opening to be fixed and extend up to the ridge.

The next step is to construct the roof, including the window frame, as the work proceeds. The window frame supports the 3 in. by 2 in. plates which, in turn, support the roof timbers. The plates are allowed to project beyond the window frame so as to obtain an overhang in front of the window. When the roof has been covered, the flooring in the roof space can be fixed to enable the studding forming the sides and cheeks of the dormer to be fixed.

Fig. 8 is a cross-section through one of the dormer cheeks, and shows that the 2 in. by 2 in. studs forming the side cheeks have been kept back flush with the inside edge of the roof plate, enabling 1 in. weatherboarding to be nailed below the plate on the outside. An eaves closing board is similar to a fascia, and is cut to fit round the roof rafters, so closing the opening between rafters at the eaves. Fig. 9 gives details around one of the jambs of the window frame. The frame material is from 4 in. by 3 in. The weatherboarding is allowed to overlap the ends of the window frame, and is finished with a cover fillet and quadrant.

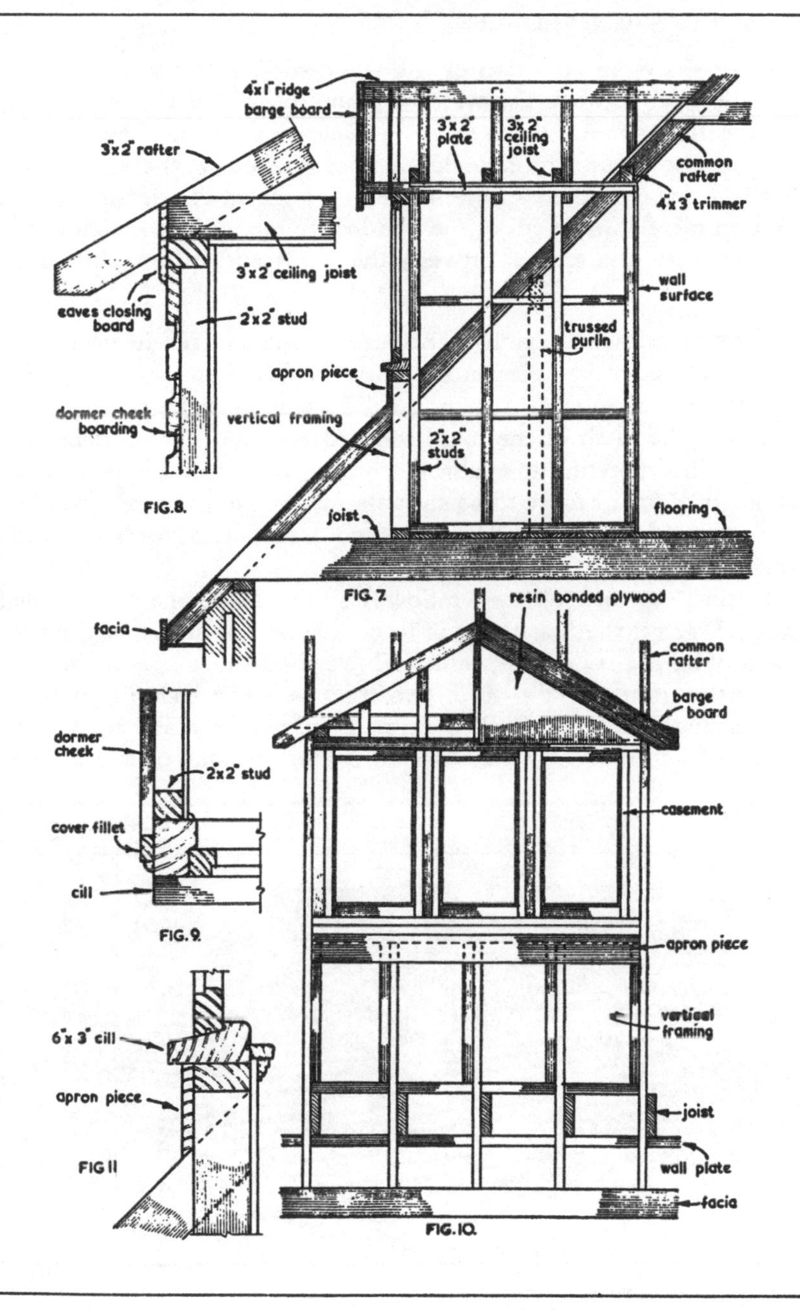
4"x1" ridge
barge board
3"x2" rafter
3"x2" plate
3"x2" ceiling joist
common rafter
4"x3" trimmer
3"x2" ceiling joist
eaves closing board
2"x2" stud
wall surface
trussed purlin
apron piece
dormer cheek boarding
vertical framing
2"x2" studs
FIG.8.
joist
flooring
FIG. 7.
resin bonded plywood
facia
common rafter
barge board
dormer cheek
2"x2" stud
casement
cover fillet
cill
FIG.9.
apron piece
6"x 3" cill
vertical framing
apron piece
joist
FIG 11
wall plate
facia
FIG.10.

Fig. 10 shows an elevation of the woodwork. Short 2 in. by 2 in. studs are fixed between the window frame head and the roof slopes to give a fixing for ¼ in. thick resin-bonded plywood which closes the space above the window. Barge boards fixed to the end rafters will give a finish to the appearance. Fig. 11 shows the details around the apron piece and the sill of the window frame. Before finishing the inside surfaces, the spaces between the studs and rafters should be well insulated.

Eyebrow window. These have become popular in recent years, but lack of knowledge has prevented many people from including this attractive feature in their own houses. First it must be realized that the steeper the pitch of the main roof rafters, the more suitable the roof is for incorporating an eyebrow window. A roof should be pitched at an angle of 60° or more to be suitable for the inclusion of a window of this type, though they have been constructed in roofs of much lower pitch.

Assuming that an eyebrow window has to be made in a roof pitched at 52½°, Fig. 12, the first thing is to decide on the shape of the eyebrow profile, Fig. 13. This profile, which gives the shape of the eyebrow at the front, is built up in two thicknesses in a similar manner to the shaped rib of an arch centre. This can be made up in two halves and joined together on site. The shape of this profile is most

A view of an eyebrow window.

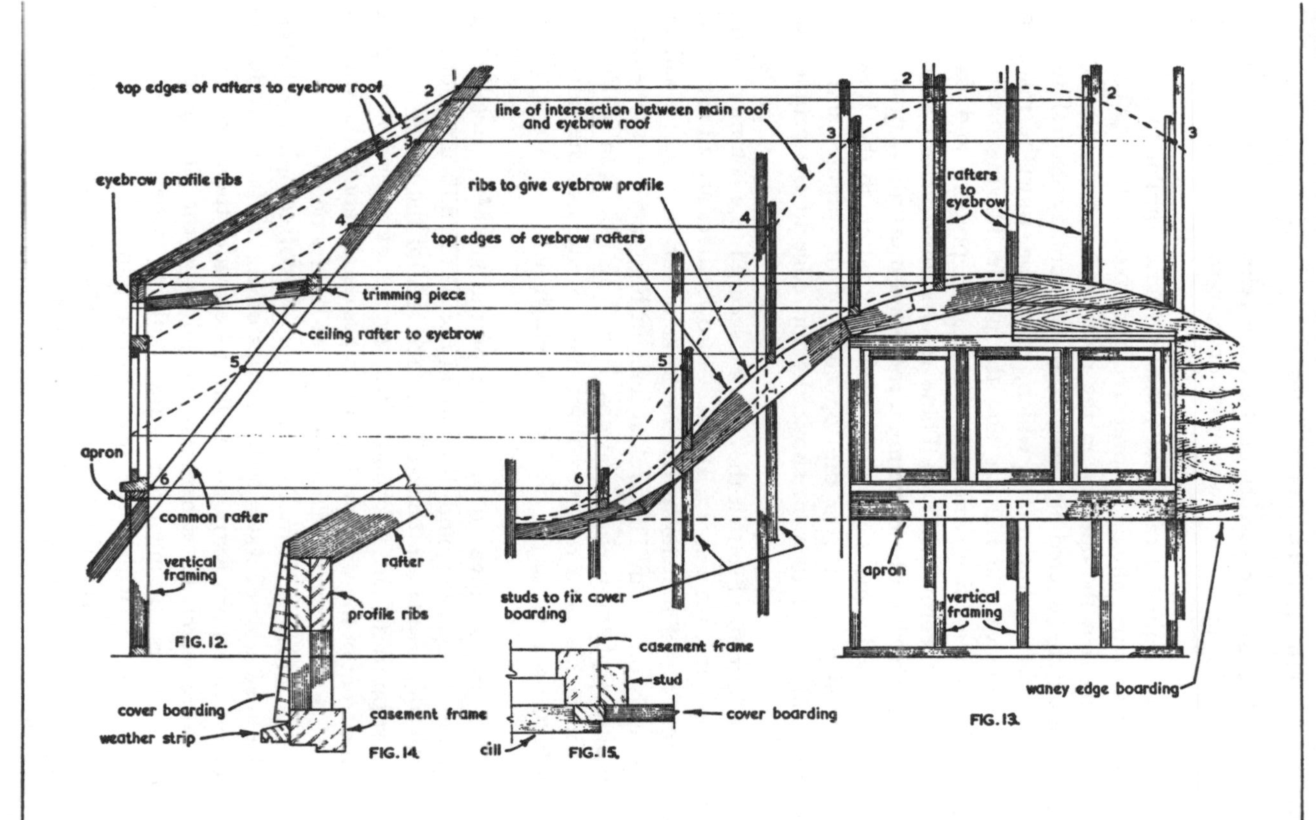

FIG.12. FIG.14. FIG.15. FIG.13.

important, because the success of the work relies on its appearance. One should aim at a low, sweeping curve, rather than a high and rounded shape.

A vertical framing to carry the window frame, similar to that for the dormer window, should be prepared and fixed as in Figs. 12 and 13. Assuming that the opening in the roof has been completed, the profile should be offered up and fixed temporarily central to the position of the window frame, which, of course, must be made to fit between the vertical framing and the profile. Having positioned and fixed the window frame and the profile, prepare and fix the rafters to the eyebrow window. These rafters must all run from the profile board, backwards to the main roof, and must all be pitched at a pre-determined angle, say 30°. The eyebrow rafters in the drawings have all been notched over the top edge of the profile, and their edges are all, say, 1 to 2 in. above the top edge of the profile, see Fig. 13.

They are also fixed so that they can be secured to a common rafter at their top ends, see Fig. 13. If the eyebrow rafters are all set at the same angle the shape shown by the broken line in Fig. 13 will be automatically obtained. The broken lines in Fig. 12 show the lengths and positions of the eyebrow rafters. It will also be seen in Fig. 13 that a corner has been removed from the top edges of the rafters to give a better seating to the tile battens which flow over the eyebrow roof.

Figs. 12 and 13 give an indication of the geometry involved in setting out the eyebrow, and this should not be too difficult to understand. Vertical studs can be fixed on each side of the window frame to give support to the plywood, tiles, or waney edge-boarding to finish the front of the window. Fig. 14 shows details above the window frame and Fig. 15 the details at the side of the frame.

There is difficulty in making an eyebrow roof completely waterproof because a lot of water which settles on the surface will flow in a sideways direction. Extra efforts must be made, therefore, to stop the moisture which flows between the joints in the tiles getting to the roof timbers. Bitumen felt and/or building paper is often used for this purpose, with large overlapping joints, but it must be emphasized that this is the weakness of the eyebrow window. Finally, one last rule; keep the rafters to the eyebrow roof pitched as steeply as possible.

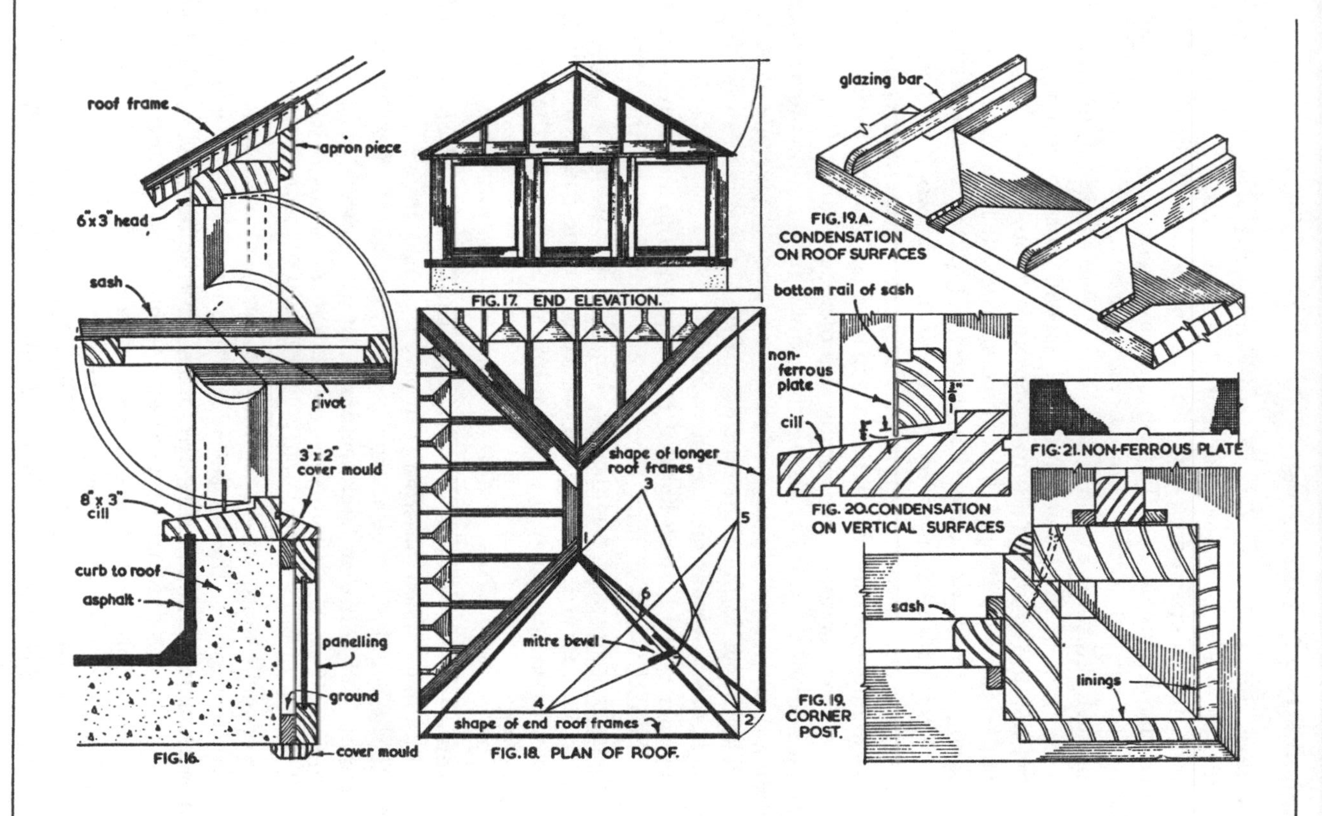

FIG. 16.

FIG. 17. END ELEVATION.

FIG. 18. PLAN OF ROOF.

FIG. 19. A. CONDENSATION ON ROOF SURFACES

FIG. 20. CONDENSATION ON VERTICAL SURFACES

FIG: 21. NON-FERROUS PLATE

FIG. 19. CORNER POST.

Lantern lights. These, Fig. 16, are constructed to give light and ventilation to a space below a flat roof. The roof, usually of reinforced concrete, has an upstanding curb around an opening in the roof. The lantern light rests on this. Fig. 17 shows an end view of a lantern light. The vertical section through the light shows that the sill sits on the curb and overhangs at least $1\frac{1}{2}$ in. so that a drip can be worked near the front of the lower edge, and asphalt worked into a groove behind the drip to ensure a fully waterproofed job.

Four frames are prepared to form the sides of the light. One method of joining them at their corners is shown in Fig. 19. The ends of the sills are mitred and secured by means of one or two $\frac{3}{8}$ in. handrail bolts with one or two $\frac{3}{8}$ in. dowels. The jambs of the frames can be recessed into the sills and rebated together as shown. Screws and glue are used for fixing the jambs together down their length, and screws and glue again for fixing the jambs to the sill, the former passing through from the back edges of the jambs at an angle of about 60 . The linings forming the outside corner of the posts can be fixed to the jambs by means of glue and pins.

The roof of the lantern light, if it is hipped as in the drawings, is made in the shape of four frames mitred together at their intersections. A method of finding the shapes of the frames is given in Figs. 17 and 18 as is the method for constructing the mitre bevels. Those who do not follow these diagrams should refer to the chapter on roofing.

Turrets. These structures, Fig. 22, are usually situated on roofs of large buildings. Their function is to give a certain amount of ventilation to the building. Some are simple in construction, others are more elaborate and form an architectural feature as well as providing ventilation. Basically, a turret consists of four posts which pass through the surfaces of the roof down to some means of anchorage. In the example shown in Fig. 22 the posts are secured to beams which run between two of the trusses supporting the roof. Above the roof surfaces, and up to the sills, the posts have infillings of studs and nogging pieces, and can be covered internally and externally with resin-bonded plywood. The outside surfaces would later be covered with lead, copper or some other sheet metal.

Above the plywood surfaces are four frames with louvre boards, and these frames have been recessed into the sides of the four main

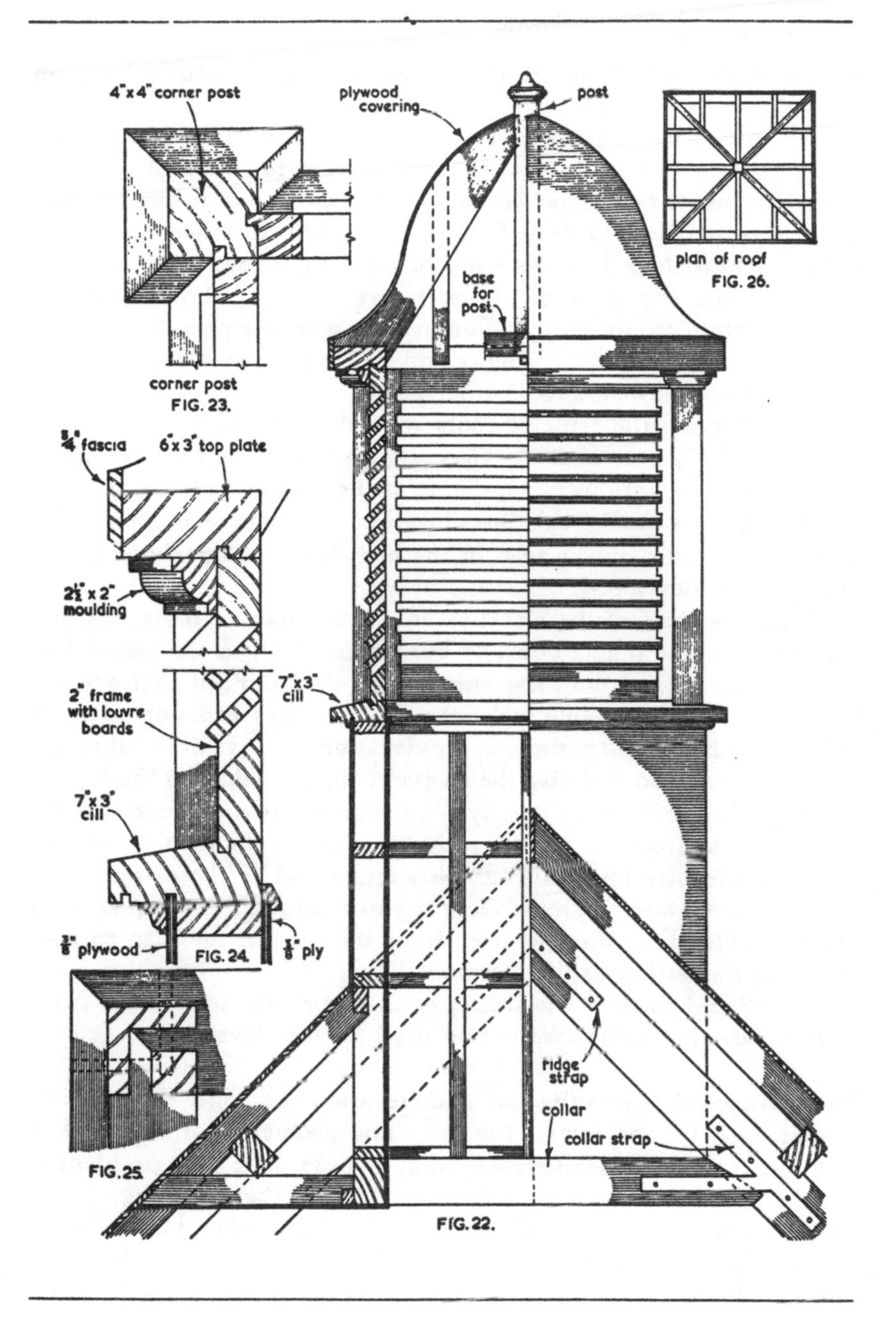
4"x4" corner post
corner post
FIG. 23.
¾" fascia
6"x3" top plate
2½" x 2" moulding
2" frame with louvre boards
7"x3" cill
⅜" plywood
FIG. 24.
⅜" ply
FIG.25.
plywood covering
post
base for post
7"x3" cill
ridge strap
collar
collar strap
FIG. 22.
plan of roof
FIG. 26.

posts. On the left of Fig. 22 is a section through the turret, and on the right an elevation.

Fig. 26 is a plan of the turret roof and is simple in design. The centre post of the roof is secured at the bottom to a base plate which extends across to two of the top plates. The centre post passes through the centre of the turret roof, continues some distance above the surface, and is shaped to give a finish. Figs. 23, 24 and 25 give larger details of the construction of the turret. Fig. 23 shows how the louvred frames are tongued and grooved to a corner post, Fig. 24 is a section through one of the frames and Fig. 25 shows a corner of the sill mitred and dowelled to a corner post.

Before leaving the topic of roofs with shaped ribs, two examples are given in Figs. 27 and 30. They indicate methods for developing the shapes of the various parts. Fig. 27 is the plan and section through a roof which is semicircular in elevation and square in plan. The shapes of the ribs can be seen in the elevation but the shape of the hips must be developed.

Divide one half of the elevation into a number of parts, say six, and project these points down to the plan and on to the centre line of one of the hips. From these points draw lines at right angles to the hip, and from a base line make the various lines brought up from the hip equal in length to those in the elevation. A curve drawn through the points obtained will give the shape of the top edges of the hips.

To obtain the shape of the plywood required to cover one surface, draw lines out from the points on the two hips in the plan, as shown, and on the centre line mark off distances equal to those round the half of the elevation. Draw vertical lines through these points to intersect with those brought over from the hips in the plan to give points on the outline of the plywood shape.

Figs. 28 and 29 show the shapes of the short ribs seen in the plan. In each case the radius used is that used for the elevation.

Ogee shape. Fig. 30 is the plan and elevation of an ogee roof, similar to that used for the turret, Fig. 22. The geometry for this roof is very like that of the semicircular roof, Fig. 27. The only problem is how to mark the backing on the top edge of the hips, Fig. 31. A templet is made exactly the shape of the hips. It is placed on each hip in turn and allowed to slide sideways the distance x in the plan, Fig. 30. The outline of the templet will give the outline of the backing.

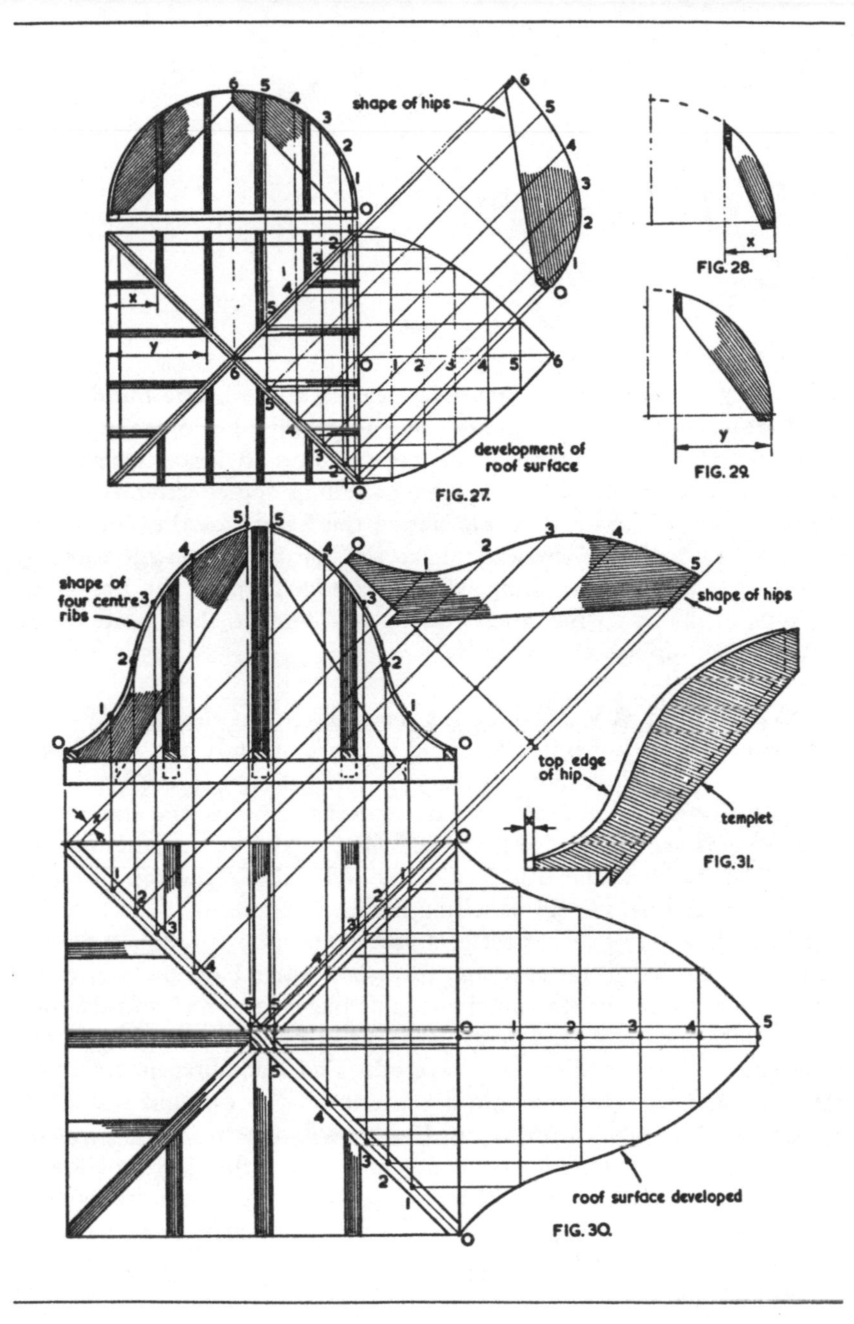

FIG. 27.
FIG. 28.
FIG. 29.
FIG. 30.
FIG. 31.

14 Panelling to Walls

Although wall panelling is fast disappearing in domestic buildings a great deal is still being carried out in public buildings such as magistrates courts, town halls, director's suites in large commercial buildings, and so on. Most of the panelling today extends up to frieze height, which is anything above the height of the doorways, or to the full height of the walls up to the cornice. Years ago most of this work was dado panelling, which extended up to the height of the tops of chairs so that these would not damage the plaster work when pushed up against the walls.

Dado panelling. A vertical section through and an elevation of some dado panelling is shown in Fig. 1 and is approximately 3 ft. 6 in. high. It consists of panelled frames to any required design secured to wood grounds carefully fixed to the wall in vertical and horizontal alignment. The grounds are usually fixed to wood plugs which have been inserted in vertical joints of the brickwork, the walls above the top grounds being plastered before the panelling is fixed.

Fig. 2 shows, to a larger scale, a vertical section through the dado panelling, and it will be seen that the wood plugs have been inserted into the brickwork joints and their front edges trimmed so that they are in perfect alignment. The grounds are fixed to the plugs, and the wall above the panelling is plastered. The panelling is fixed by screwing into the grounds, and a capping on the top and a skirting at floor level completes the work. If a moulded strip is first screwed to the floor and the tongue on the bottom edge of the skirting allowed to fit into a groove on the moulded strip, the skirting need be fixed at the top only, so allowing movement to take place in the timber without creating an open joint between the bottom of the skirting and the floorboards.

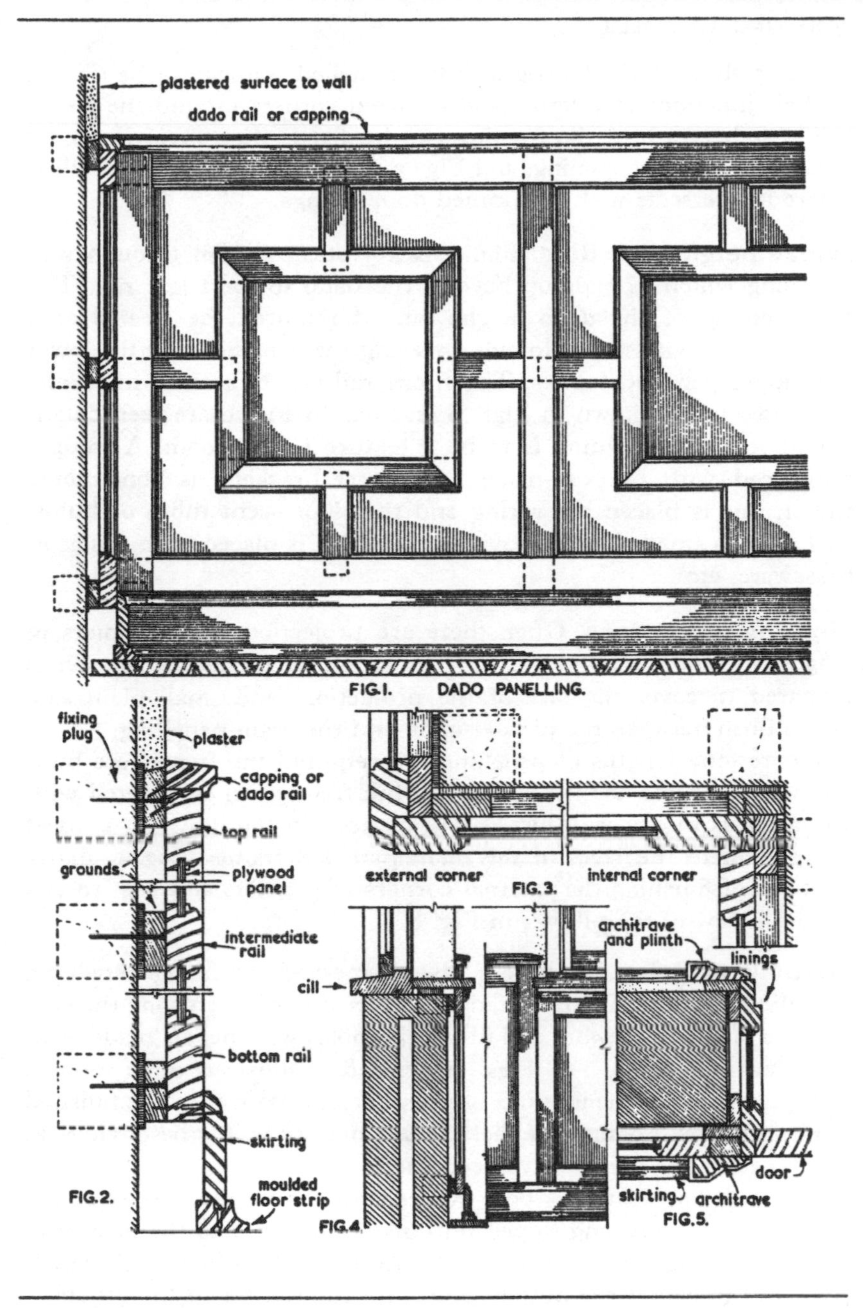

FIG.1. DADO PANELLING.

Fig. 3 shows how the edges of the panelled frames can be treated at their junctions at internal and external corners around the room. Fig. 4 indicates how the framing can be fixed when involving a sill board at a window opening, and Fig. 5 gives details of dado panelling where it intersects with the framed door linings.

Frieze height panelling. Fig. 6 is a vertical section through wall panelling which extends up beyond the dado to the frieze rail. The panelling up to the dado height can, if required, be treated as a separate unit with the dado rails covering the gaps between the lower and upper panelled frames. The frieze rail can be treated in several ways, two being shown in Figs. 6 and 6a. In Fig. 6 are seen details where concealed lighting is to be a feature in the room. A trough, reinforced with purpose-made mild steel brackets, is constructed and in this is placed the wiring and the fluorescent tubes or bulbs. In Fig. 6a a small shelf is provided on which is placed such things as brassware, etc.

Clearing projections. Often there are projections such as piers in long walls, and Fig. 7 shows how these are negotiated. A pilaster is prepared to cover the face of the projection, and small plain side pieces fit in between the pilaster ends and the main panelling.

Where long lengths of panelling are required the frames are built up in sections and the joints between the frames can be covered with a pilaster as in Fig. 8. This, as will be seen in the drawing, is fixed straight on to the faces of the main panelled frames. Fig. 9 shows methods of finishing the external corners of pilasters and Fig. 10 is a pictorial view of the pilaster in Fig. 8.

Methods of fixing. There are several methods of fixing panelling, including pilasters. Often the fixing is carried out by screwing through the face of the framework and filling the holes with pellets made from a similar timber, Fig. 11a. Figs. 11, 12 and 13 illustrate other ways of fixing panelling to the grounds used in better class work. The finished work will show no signs of the fixings; neither will it be obvious at which points the fixings have been used.

In Fig. 11 a rebate has been worked on the top back edge of the horizontal grounds, and lipped buttons are screwed to the back surfaces of the panelled frames. The frame is fixed by 'hanging', which involves the lips of the buttons engaging the rebates in the grounds.

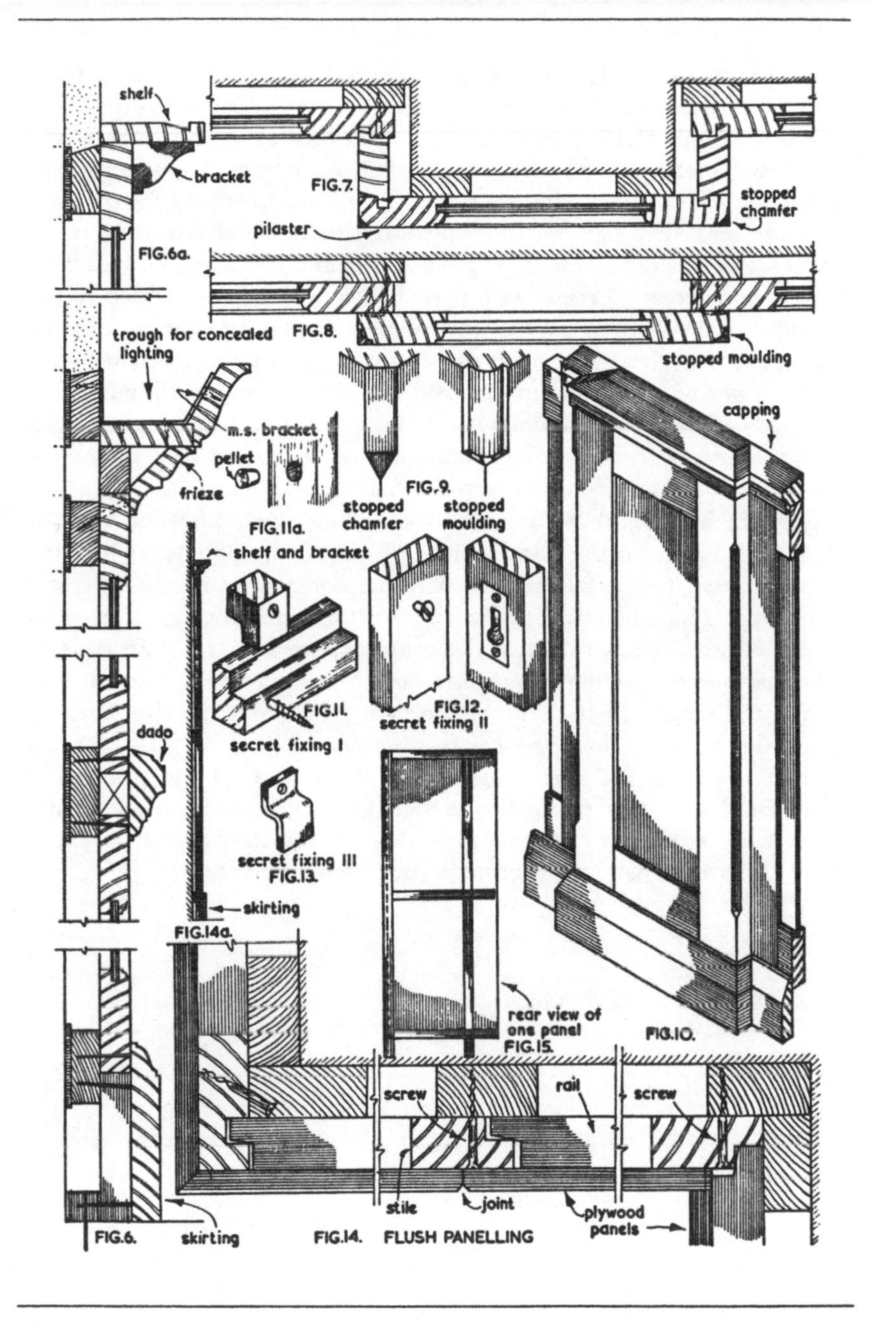
shelf
bracket
FIG.6a.
FIG.7.
pilaster
stopped chamfer
FIG.8.
trough for concealed lighting
stopped moulding
m.s. bracket
pellet
frieze
FIG.11a.
FIG.9.
stopped chamfer
stopped moulding
capping
shelf and bracket
FIG.11.
secret fixing I
FIG.12.
secret fixing II
dado
secret fixing III
FIG.13.
skirting
FIG.14a.
rear view of one panel
FIG.15.
FIG.10.
screw
rail
screw
stile
joint
plywood panels
FIG.6.
skirting
FIG.14. FLUSH PANELLING

When in position the frames are secured to prevent them from becoming dislodged from the grounds. This is done by inserting screws through the frames into the grounds at points where the screws will not be seen, for instance, behind where the skirting will be fixed, and at the top where the screws will be hidden by the capping or frieze rail. Instead of wood buttons being used the mild steel button, Fig. 13, can be used. Fig. 12 shows another method, based on the slot and screw. Screws, with their heads projecting are fixed to the grounds at convenient points and plates, similar to that in Fig. 12 are secured to the back surfaces of the panelling, to coincide with the positions of the screws in the grounds. When the panelling is fixed, the heads of the screws are allowed to enter the holes in the plates and the panelling is then carefully tapped downwards, so that the screw shanks enter the narrower slots in the plates.

Figs. 14 and 15 show a method of fixing flush plywood panels. At Fig. 15 is shown the back surface of one of the panels. It has had a frame glued to the surface, each frame consisting of one stile and three rails. The stile overlaps the edge of the panel by, say, 1 in., and is rebated on the edge nearest the grounds, see Fig. 14. The ends of the rails farthest from the stile are cut to form a tongue which will fit into the rebate on the stile of the next panel, Fig. 14. By working round the room in the appropriate direction each panel can be fixed by first inserting the three tongues in the rebate of the previously fixed panel and then fixing the other edge of the panel by screwing through its stile into the grounds. Also on the drawing, Fig. 14, is shown how external and internal corners can be treated.

15 Counter Construction

The first of two methods for constructing counters are shown in Figs. 1 and 2, which are a vertical cross-section and part elevation of what could be described as a traditional style counter. The three components are a top of solid timber, a framed and panelled front with raised and fielded panels and bolection mouldings, and the pedestal or cupboard section. This counter is suitable for a bank or showroom which has been built on the lines frequently used between the wars and before that period.

The top, of solid hardwood timber such as mahogany, is 1¼ in. thick, and this has been increased in the front by another piece 1¼ in. thick to give the appearance of its having been prepared from much heavier timber, see Fig. 5. The panelled front is made up in sections (depending on the overall length of the counter) with shaped, solid pilasters covering the joints between the panelled sections. The cupboard or pedestal sections can be made up as separate units, and can be arranged throughout the length of the counter as desired.

Fig. 6 is a pictorial view of a pedestal and comprises drawer and cupboard spaces. These units can be made from softwood and plywood, and painted or stained to match the hardwood top and front. The spaces between the units can be utilised as necessary. If clerks are to sit behind the counter, as in banks, these are adapted as knee spaces.

Modern style counter. Fig. 4 is a vertical section through a more modern type of counter, the top and front being made almost entirely from blockboard. The joints between the various panels which form the front of the counter can be masked by plain pilasters and the top built up at its front edge as in Fig. 7. The pedestal, which is also

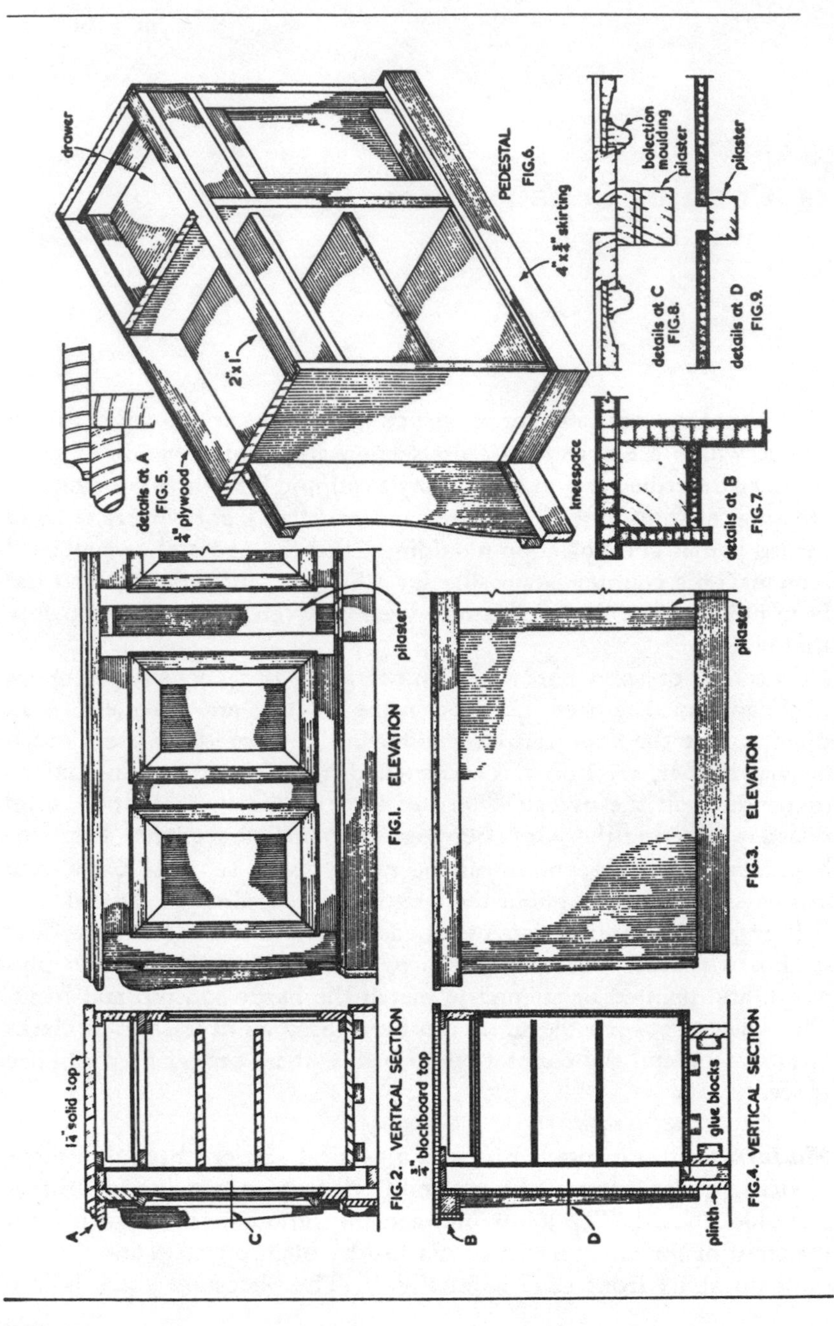

FIG.1. ELEVATION

FIG.2. VERTICAL SECTION

FIG.3. ELEVATION

FIG.4. VERTICAL SECTION

details at A FIG.5.

PEDESTAL FIG.6.

details at B FIG.7.

details at C FIG.8.

details at D FIG.9.

made from blockboard, sits on a plinth frame made from $1\frac{1}{2}$ in. timber strengthened by glue blocks.

Figs. 8 and 9 are cross-sections through the pilasters in Figs. 2 and 4. The top and front of the blockboard counter are often surfaced with a plastic such as *Formica,* and if the surfaces of the pilasters are kept flat, these, as well as the skirting, can also be covered with the same or a contrasting plastic.

16 Church Furniture

Some of the best examples of craftsmanship can be seen in our churches and cathedrals. Even today, when much woodwork appears drab and uninteresting, the quality of craftsmanship in churches seems to stand out from all other work. Quite plain church joinery can be pleasing to look at if carefully designed and carried out by competent craftsmen.

Pews. Fig. 1 shows details of church pews. These are usually placed at 3 ft. centres, and the pew ends should be about 18 in. to 20 in. wide. The end, Fig. 1c, is a frame of two stiles and two rails, all from 2 in. material, and a panel of the same thickness finishing flush with the framework on both sides. The stuck mouldings around the inside edges of the frame have mason's mitre at the corners, and the front and back corners and the frame are stop-chamfered. Stiles have $\frac{3}{4}$ in. thick tenons at their lower ends, these entering into mortices prepared in the 3 in. by 3 in. curb. The joints are secured with hardwood dowels. A suitably moulded capping is fixed along the top edge of each pew end and secured with glue and dowels.

To support the floor to the pews, rebates are worked on the top inside edges of the curbs. In addition, 2 in. by 2 in. bearers are placed at about 16 in. centres along the length of the pews to the curb at the opposite end, see Fig. 3. These give support to the floor boards.

Fig. 1b gives details of the seating to the pews, comprising a panelled and framed back, and a $1\frac{1}{2}$ in. thick seat resting on supports spaced at approximately 4 ft. centres.

A book rest is provided for hymn book and bibles for the pew behind, and a hassock rest is arranged below the seat. At Fig. 1a are details of the front pew screen which consists of a panelled frame

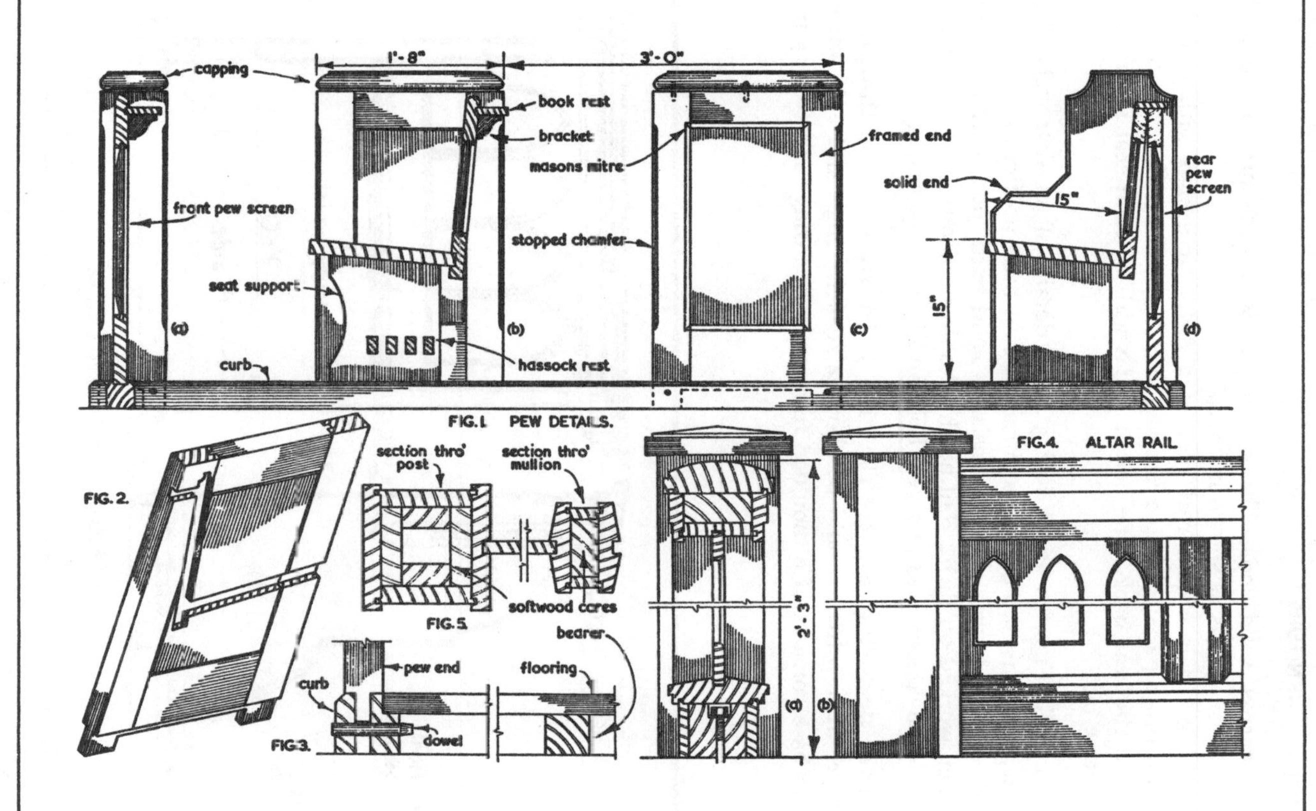
1'-8"
3'-0"
capping
book rest
bracket
masons mitre
framed end
front pew screen
solid end
15"
rear pew screen
stopped chamfer
seat support
15"
(a)
(b)
(c)
(d)
hassock rest
curb
FIG. 1 PEW DETAILS.
FIG. 2.
section thro' post
section thro' mullion
softwood cores
FIG. 5
pew end
curb
dowel
flooring
bearer
FIG. 3.
FIG. 4. ALTAR RAIL
2'-3"
(a)
(b)

with narrow solid ends. The seat and back of each pew is recessed into the pew ends as shown in Fig. 2.

At Fig. 1d is seen a rear pew of a different design with screen, comprising a panelled frame recessed into the return curb, immediately behind the seat back. The height and width of each seat should be about 15 in.

Altar rail. Figs. 4a and b and 5 show details of an altar rail. The posts, mullions, top and bottom rails are all built up from hardwoods with softwood cores. Panels consist of solid hardwood with a tracery design worked on them. Rail height should be about 2 ft. 3 in. and it should be constructed to give a bold and heavy appearance.

Prayer desk. Fig. 6 is the elevation and vertical section through a prayer desk, which should be about 2 ft. 6 in. high and 2 ft. wide. The base consists of a moulded frame and boarding, and two 1½ in.

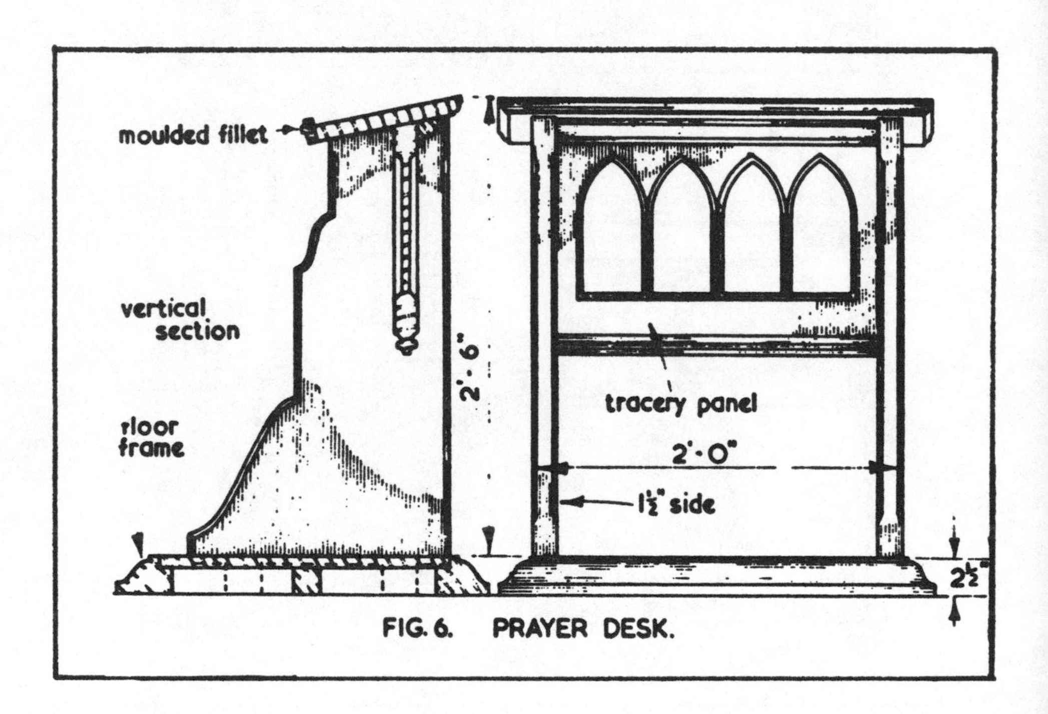

FIG. 6. PRAYER DESK.

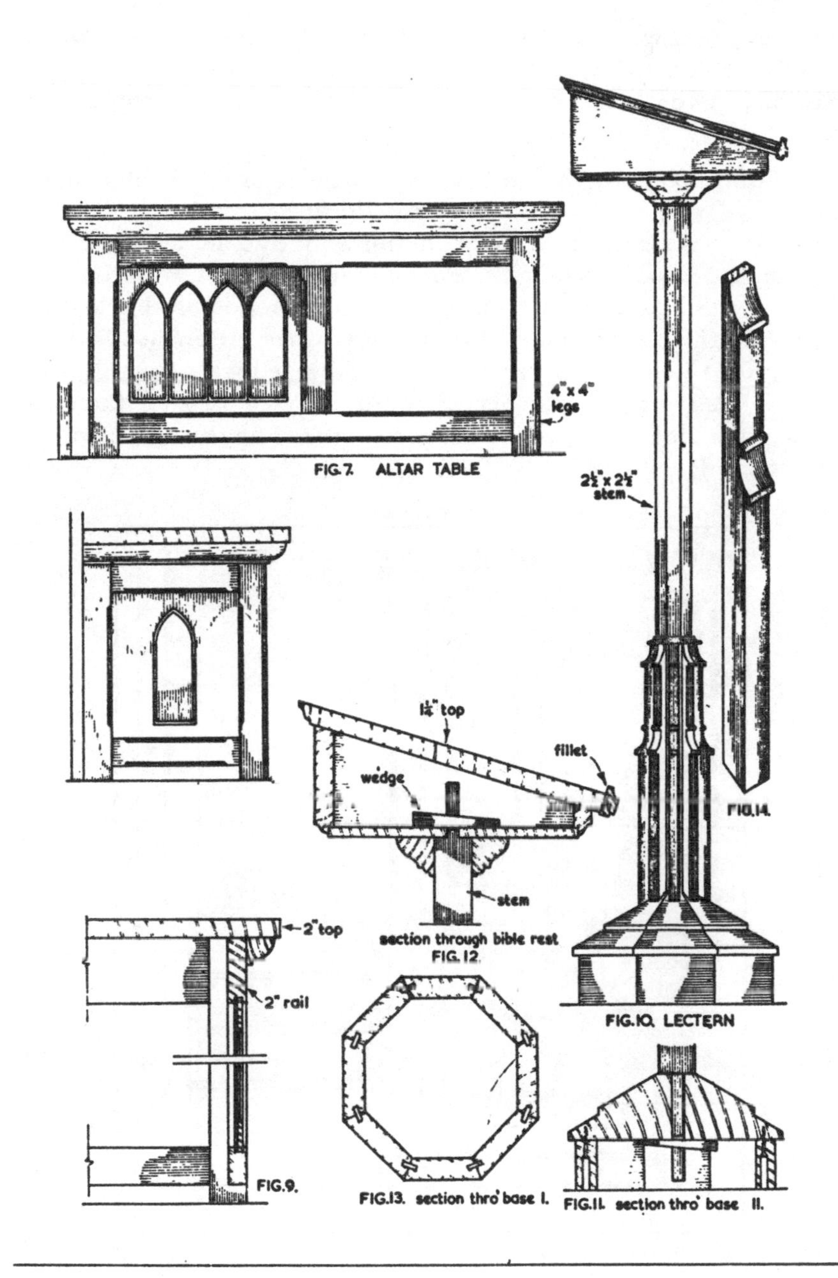

FIG. 7. ALTAR TABLE

section through bible rest FIG. 12.

FIG. 10. LECTERN

FIG. 9.

FIG. 13. section thro' base I.

FIG. 11. section thro' base II.

thick sides support a sloping book rest. A tracery panel can be included below the book rest if required. A moulded fillet should be included at the back edge of the book rest to prevent books from sliding off.

Altar table. Figs. 7, 8 and 9 show details of an altar table and a glance at Fig. 9 indicates how this type of fitting can be constructed. The table has a 2 in. thick top with four 4 in. by 4 in. legs, and 8 in. by 2 in. and 4 in. by 2 in. rails. The rails are kept flush with the front edges of the legs. A bold 3 in. by 3 in. moulding is fixed below the top round three sides of the table. Tracery panels are framed between the legs and rails. These panels should be backed with ¼ in. plywood panels, faced with a wood veneer similar to that used for the main parts of the table. Stopped chamfers are worked on the corners of the legs and rails.

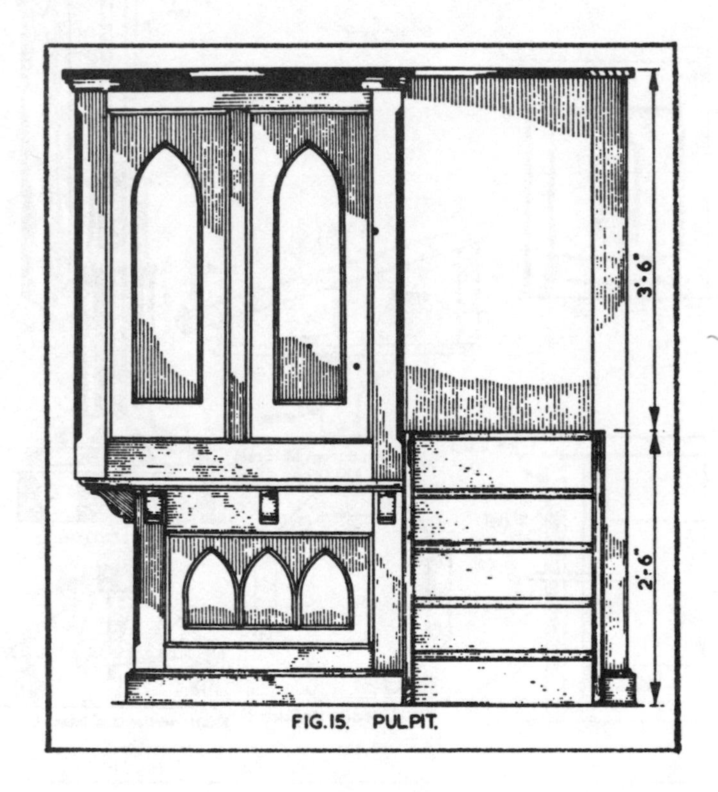

FIG.15. PULPIT.

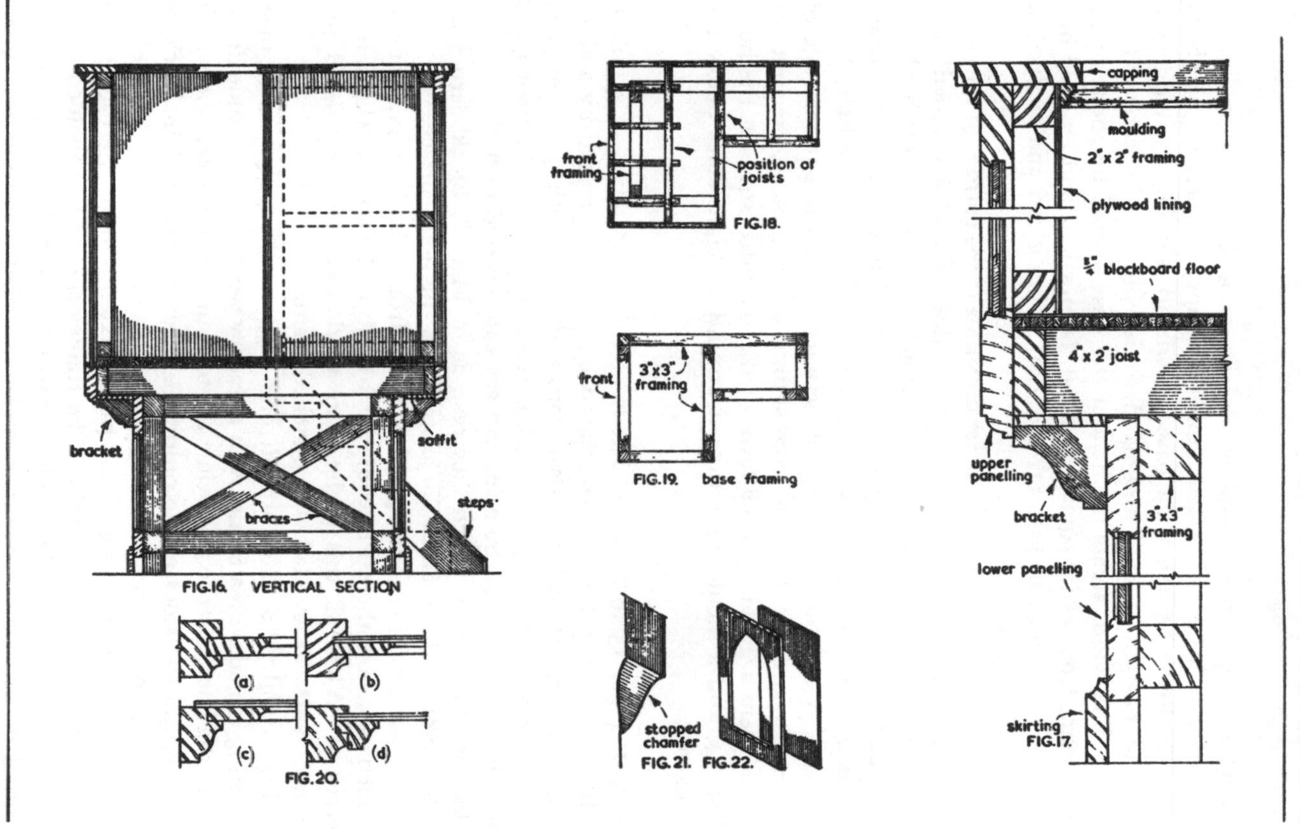
capping
moulding
2" x 2" framing
plywood lining
5/8" blockboard floor
4" x 2" joist
3" x 3" framing
upper panelling
bracket
lower panelling
skirting
FIG.17.
position of joists
front framing
FIG.18.
3" x 3" framing
front
FIG.19. base framing
stopped chamfer
FIG.21. FIG.22.
soffit
steps
braces
bracket
FIG.16. VERTICAL SECTION
(a)
(b)
(c)
(d)
FIG.20.

Lectern. Figs. 10 to 14 show details of a lectern. This has an octagonal built-up base, a stem similar in shape, and a rectangular book rest. The lower portion of the base is made up as in Fig. 13, and its top stepped portion is solid and fixed to its plinth with screws from below. The width across the flats of the base should be about 15 in. The stem, from 2½ in. square material, is morticed and tenoned into the base, as shown in Fig. 11, the joint being glued and wedged as seen in the details.

Bible rest construction has secret dovetails at the corners of the sides. The base should be rebated into the sides, pinned and glued. The tenon at the top of the stem passes through a mortice in the book rest base and is glued and wedged in position. Fixing of the top surface of the bible rest is with screws through into the sides, the holes being pelleted. A moulded fillet should be fixed along the bottom edge of the rest to prevent the bible from sliding to the ground. Decorations to the stem are added in the form of the moulded strips at the base, Fig. 14, and a moulding is mitred around its eight sides below the book rest.

Pulpit. Figs. 15 to 22 show details of the construction of a pulpit. The floor is 2 ft. 6 in. above ground level and is reached by a short flight of steps at the rear. Base construction is with 3 in. by 3 in. braced framing, the plan of which is seen in Fig. 19. The floor joists to the pulpit are 4 in. by 2 in. and are cantilevered over the sides of the base framing on which they rest, as in Fig. 18. Blockboard ¾ in. thick screwed down on to the joists forms the floor surface.

The sub-framing of the pulpit sides is from 2 in. by 2 in. material which rests on the surface of the blockboard floor to which it is screwed. When all this preliminary work has been completed the outside surfaces are clad with panelled frames.

Fig. 17 gives details through one edge of the pulpit. The base framing is clad with a frame, with a tracery panel which extends down from the floor joists to just below the top surface of the skirting which runs round the sides of the pulpit. A hardwood soffit, ½ in. thick, is fixed to the lower surfaces of the joists with 2 in. shaped brackets below.

Wall panelled framing of the pulpit extends from the capping at the top to just below the soffit. The frames should be moulded along their bottom edges. The inside surfaces of the walls can be covered with ¼ in. plywood. Mouldings can be returned round the tops of the inside and outside surfaces of the pulpit walls just below the hardwood capping. Figs. 20 and 22 show details of methods of forming tracery panels. At 20a is a solid panel, at 20b is shown a solid panel backed with a plywood, both fitting into a groove worked in the edge of the framing, 20c shows a similar arrangement but here both pieces are fixed to the back edge of the framing. 20d shows a more elaborate form of panel backed with plywood.

Fig. 22 is a pictorial view of a tracery panel with its plywood backing, and at Fig. 21 are details of a stopped chamfer which can be worked on the edges of the framework. In Figs. 15 and 16 can be seen details of the flight of steps leading up to the pulpit.

17 The Construction of Stairs and Handrails

Straight flights, dog-legged stairs, and open-newel stairs have already been dealt with in the companion book, *Practical Carpentry and Joinery*. The types to be described in this chapter are known as geometrical stairs. These do not contain newel posts and therefore have to be constructed with continuous strings and handrails.

Details of a geometrical staircase. Let us first consider the general layout of a geometrical stair. Fig. 1 is the plan of such a stair, and comprises two straight flights of five steps in each, these flights being connected by three winders where the stairs change direction. The first step in a traditional geometrical stair is usually shaped like the one shown on the plan which is called a curtail step. The shaped end generally follows the curve of the handrail scroll which is immediately above it.

A variation to the curtail step is the commode step, the outline of which can be seen in broken line. This is constructed in the same way as the curtail step but it also has a curved riser which goes across the full width of the stairs. The plan of the stairs is only one of many layouts that it is possible to get with geometrical staircases. For instance, the two straight flights can be connected with a quarter-space landing instead of the winders; the turn of the stairs can be 180°, not 90° as shown; also the two parallel flights can be connected by a half-space landing; or a half-turn of winders, or even a quarter-space landing and a quarter-turn of winders.

String details. The wall strings are constructed similarly to those for dog-legged and open-newel stairs, but the outer strings are often cut to the outline of the steps, as seen in Figs. 2 and 3. These cut strings, as they are called, are prepared so that the treads actually rest on the top horizontal surface of the step outline, and the risers are

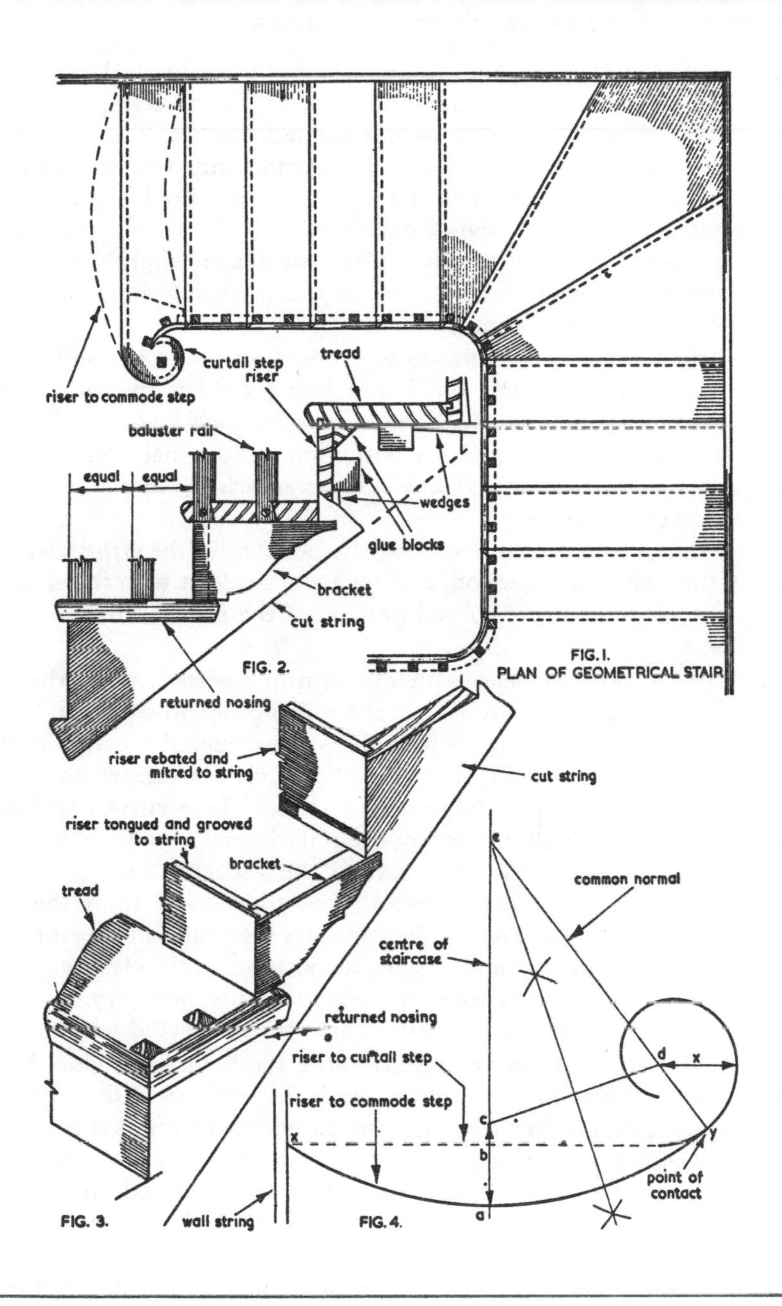
riser to commode step
curtail step
tread
riser
baluster rail
equal
equal
wedges
glue blocks
bracket
cut string
FIG. 2.
returned nosing
FIG. I.
PLAN OF GEOMETRICAL STAIR
riser rebated and
mitred to string
cut string
riser tongued and grooved
to string
bracket
tread
centre of
staircase
common normal
returned nosing
riser to curtail step
riser to commode step
point of
contact
FIG. 3.
wall string
FIG. 4.

either mitred or tongued-and-grooved to the front vertical edge, see Fig. 3. If the risers are tongued-and-grooved to the string, the end grain of the riser is concealed with a shaped bracket. This is a thin piece of solid timber or plywood, glued and pinned to the surface of the string. No bracket is required if the riser is mitred to the string.

The top step in Fig. 2 shows details of how the treads and risers are tongued-and-grooved together. These details are slightly different from those for steps in closed or uncut strings, being more practical for the cut-string type of stair.

The baluster rails are dovetailed to the ends of the treads and glued and screwed into position, the joints being hidden by the return nosing which is mitred to the front nosing of each step, see Fig. 3. The method used for correct positioning of the baluster rails is shown in Fig. 2. As many glue blocks as necessary should be used for strengthening the steps.

Up to now we have not mentioned the portions of the strings which involve the change of direction, neither have we dealt with the shaped steps. Let us consider the curved portion of the strings first.

Curved strings. Fig. 11 deals with the turn in the staircase at the top of the first flight, but before we consider this problem we should first take a simpler example. Fig. 9 shows how these curved portions of the strings are constructed. We have to make a drum or former on which the timber can be bent to the required shape. These curved portions of the strings are built up to any required thickness with veneers from $\frac{1}{16}$ in. to $\frac{1}{8}$ in. thick. The first of the veneers to go on to the drum should have the marking-out lines transferred from the rod to its inside surface, see Fig. 8. The veneer is then carefully positioned, bent round the drum, and fixed in position with clamps or G cramps, Fig. 9. Other veneers are added to the first, say three or four at a time, the surface of every veneer being covered with a cold-water glue such as casein or some suitable synthetic resin glue. Each veneer must be kept in close contact with the surfaces of the adjacent veneers, additional clamps being used as necessary for this purpose. When the glue has hardened the built-up string can be removed from the drum and, by following the lines which were placed on the first veneer to be placed on the drum, it can be shaped and prepared as seen in Fig. 10.

Fig. 5 shows how a continuous string for a staircase of two flights

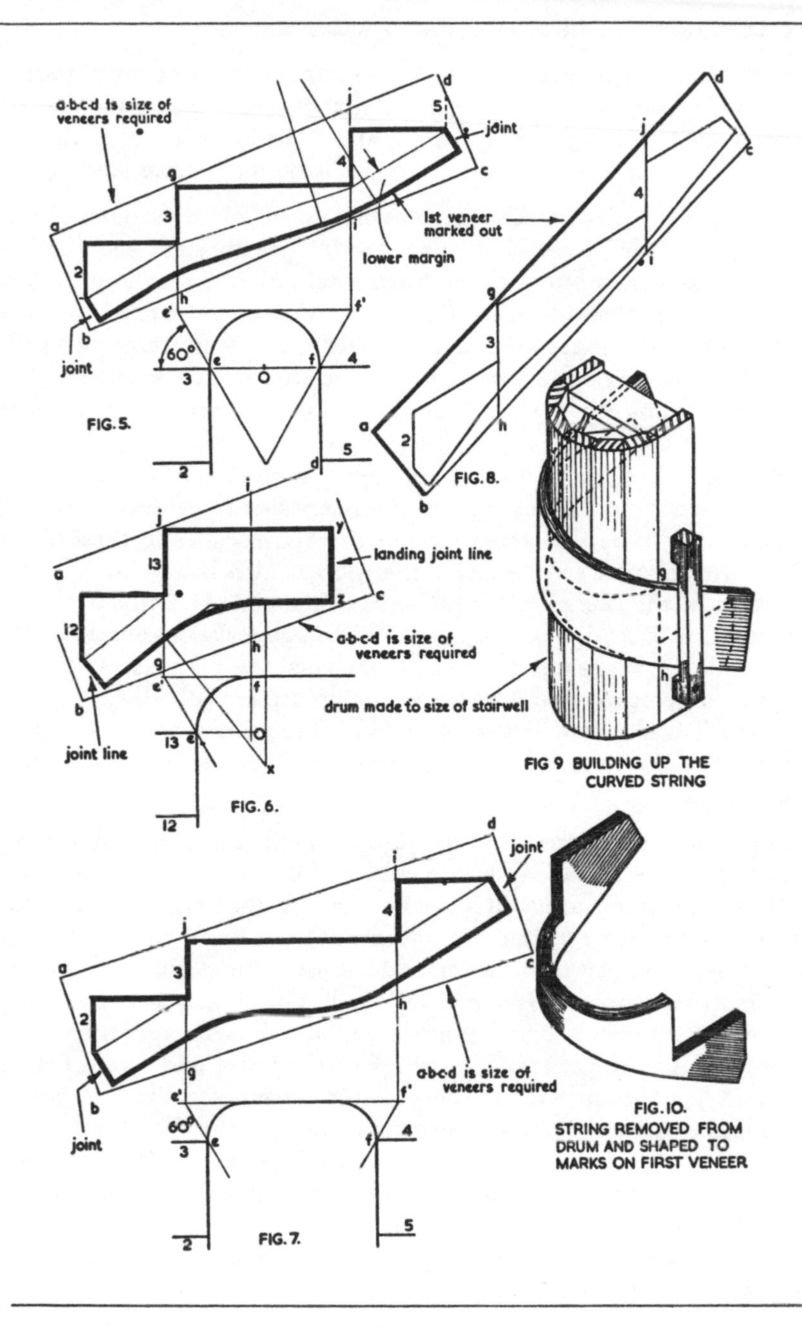

FIG. 5.

FIG. 6.

FIG. 7.

FIG. 8.

FIG 9 BUILDING UP THE CURVED STRING

FIG. 10. STRING REMOVED FROM DRUM AND SHAPED TO MARKS ON FIRST VENEER

connected by a half-space landing should be set out. The lower part of the drawing shows the plan of the stairs. Risers 2 and 3 are the top steps of the lower flight, and risers 4 and 5 the lower risers of the top flight. The space between risers 3 and 4 is the half-space landing.

Preparing a stretch-out. To obtain what is called the stretch-out of the steps, two 60° lines should be drawn, one through each end of the diameter e–f to intersect the horizontal line just touching the top of the curve in e′ and f′. The distance e′–f′ is the horizontal distance from e round to f seen in the plan. Project e′ and f′ upwards vertically so that the stretch-out of the steps can be constructed as shown. The only dimensions required to draw the stretch-out are the 'going' and the 'rise' of each step.

Having drawn the outline of the steps in the stretch-out, straight lines connecting the lower ends of the risers should be drawn and the distance known as the lower margin marked parallel to these lines. This lower margin can be any dimension, and depends on the construction of the stairs. If, for instance, the stairs are fairly wide, say 4 ft. 6 in. to 5 ft., it may be necessary to support the centres of the steps with a carriage and a series of brackets, see Fig. 14a.

As it is necessary to add a soffit to the underneath surface of the staircase it is also essential to have the lower margin wide enough to come below the lower edge of the carriage pieces so that the soffit is supported.

Joints. It will be noticed that the joint at each end of the stretch-out is one step behind the end of the curved portion. For instance, at the lower end of the marking out it can be seen that the joint is immediately below riser number 2, but the string does not begin to turn until it gets to riser number 3. At the top end the same thing occurs; the curve of the string stops at riser number 4, but the joint is immediately below riser number 5. These joints are where the section of the curved string is joined to the straight strings of the lower and upper flights.

The type of joint used is the counter cramp, see Fig. 14. Before placing the veneer which has been set out round the drum, it must be ascertained that the guide lines g–h and i–j are on the veneer. These will assist in placing the veneer on the drum in the correct position. If they coincide with the ends of the curved portion of the drum and are in the vertical direction, it is fairly safe to assume that the 'rise' or position of the veneer is correct.

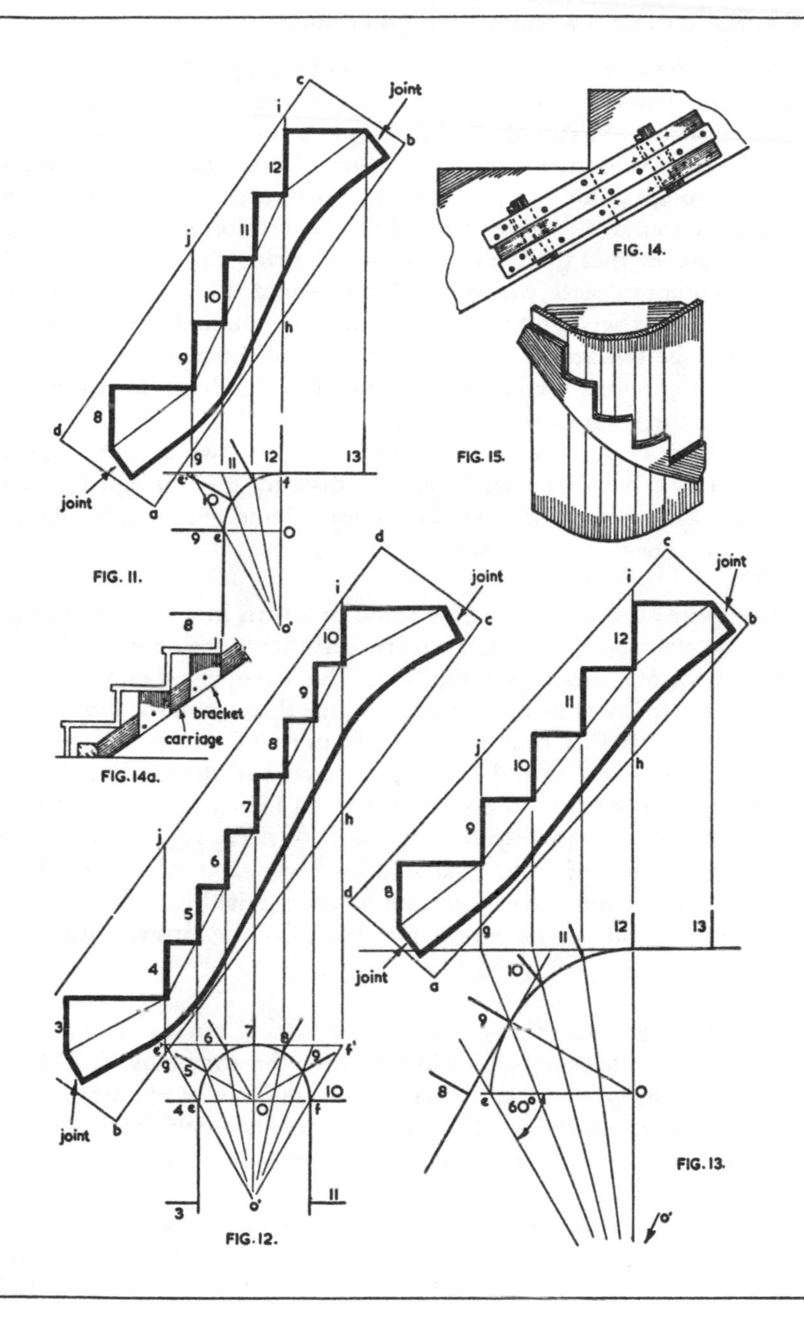
joint
joint
FIG. 11.
FIG. 14.
FIG. 15.
FIG. 14a.
bracket
carriage
FIG. 12.
FIG. 13.
60°

Fig. 10 shows what the curved portion of string will look like when it has been removed from the drum and shaped.

Fig. 6 is a plan of a straight flight rising to a landing, riser number 13 being the last one on the stair. As the turn in this example is of 90°, a 60° line through point e to intersect the horizontal line in e′ will give the distance round the curve from e to f, the developed distance being e′–f. Remember that the curved portion of string this time is jointed to the front apron piece which forms the face of the landing timbers. Therefore from riser 13 on the development, the direction of the marking out will be horizontal, following the line of the landing. The distance between points y and z will be equal to the depth of the apron piece along the front of the landing.

Fig. 7 is an example similar to that in Fig. 5, two straight flights connected by a half-space landing. The distance e′–f′ is equal to the distance e–f on the plan. Remember, lines h–i and g–j must be placed on the first veneer in every case.

Quarter space of winders. We now come to a turn in a stair similar to that in the staircase in Fig. 1. Two straight flights are connected by a quarter-space of winders (see Fig. 11). The distance e–f is developed in the usual way by drawing a 60 degrees line through e, the dimension e′–f being the distance round the turn from e to f. As we also require the positions of risers 10 and 11 on the stretch-out, the 60° line should be extended downwards to meet the vertical line from o in o′. Lines from o′ should be drawn through the ends of risers 10 and 11 to give their positions on line e′–f. It should be fairly clear what should be done from this point to complete the marking out. Fig. 15 shows the completed portion of the string put back on the drum until it is required

Half-space of winders. Fig. 12 shows the setting out for two flights connected by a half-space of winders, and the drawings should be clear enough to follow if the foregoing notes have been memorised. Fig. 13, however, shows two flights connected by a one-third space of winders, or in other words, two flights connected by winders giving the stairs a 60° turn.

The plan of the stairs should be constructed, and the curved portion can, temporarily, be changed into a 90° turn as shown. A 60° line from e downwards to meet a vertical line from o to give point o′, and lines from o′ through the ends of risers 9, 10 and 11 will give the positions of these risers in the stretch-out.

Shaped bottom step. We should now turn our attention to the shaped steps at the bottom of the staircase. Consider the curtailed step first, Fig. 20. It is necessary to know the shape of the scroll which is situated immediately above the shaped end of the step. The outline of this is first constructed (see Fig. 11 Chapter 19), and the setting out of the shaped end of the curtail step superimposed over the scroll lines, making the nosing line of the step follow the outside edge of the scroll. As can be seen from the drawings, a block is first made similar to the shape taken from the setting out (Figs. 20 and 22). This can be built up from several pieces of timber with the grain going in opposite directions like a piece of plywood. If a bandsaw is available the shape of the block is best cut on this.

A modern staircase with a centre laminated string. The treads are cantilevered on each side of the string and are fixed with purpose made brackets. The string is approximately 28 ft. long, 18 inches wide and 15 inches deep, the laminations are of 1 inch Douglas Fir and were glued together with 'casco' casein glue M1562.
Elliott, White, Reading.

The block is fixed to the end of the string by means of a tongued-and-grooved joint, and is glued and screwed. Recesses should be made in the block to receive a pair of folding wedges near the centre of the block. These are used for fixing the two veneers which meet at that point; and another recess is needed for receiving the riser less the thickness of the veneer at the front of the block.

The short veneer should first be fixed by gluing to the block, and this can be accomplished by the use of a former block and G cramps. The former block is cut to fit up against the veneer when it is in position, and the G cramps will hold the veneer tight up against the block until the glue has set.

The longer veneer is obtained by reducing one end of the first riser to between $\frac{1}{16}$ in. and $\frac{1}{8}$ in., taking care to keep the thickness uniform throughout its length. When the G cramps and block have been removed the second and longer veneer can, if necessary, be steamed or soaked in hot water to soften the fibres. The veneer and block surfaces are glued, and the end of the veneer entered into the recess for the folding wedges near the centre of the block. The wedges should then be glued and driven home, care being taken to see that the veneer is positioned correctly so that its edges will run parallel to the block when it is wrapped around into its final position.

To enable the surfaces of the block and veneer to come in to close contact with one another, folding wedges are glued and inserted between the shoulder at the end of the veneered portion and the block and carefully driven home. Screws can be used for keeping the remainder of the riser up against the block, as seen to the left of Fig. 20. The tread to the shaped step is prepared to the shape of the outline of the scroll. This is the nosing line indicated on the plan, Fig. 20.

Commode step. If this is to be constructed, the manner described for the curtail step can be followed with slight variations. The block requires two formers, one at the top and one at the bottom, recessed into it so as to form the curved riser which goes right across the width of the stairs, see Fig. 21. Saw kerfs are used in bending the thicker portion of the riser. Some difficulty, however, may be found in setting out the shape of the step, see Fig. 4. Let a–e be the centre line of the staircase, a–b the amount the curved riser is in front of its position if it were a curtail step, and a–c the radius of the last curve used when setting out the scroll. This is equal to d–y, Fig. 4. Join c and d and then

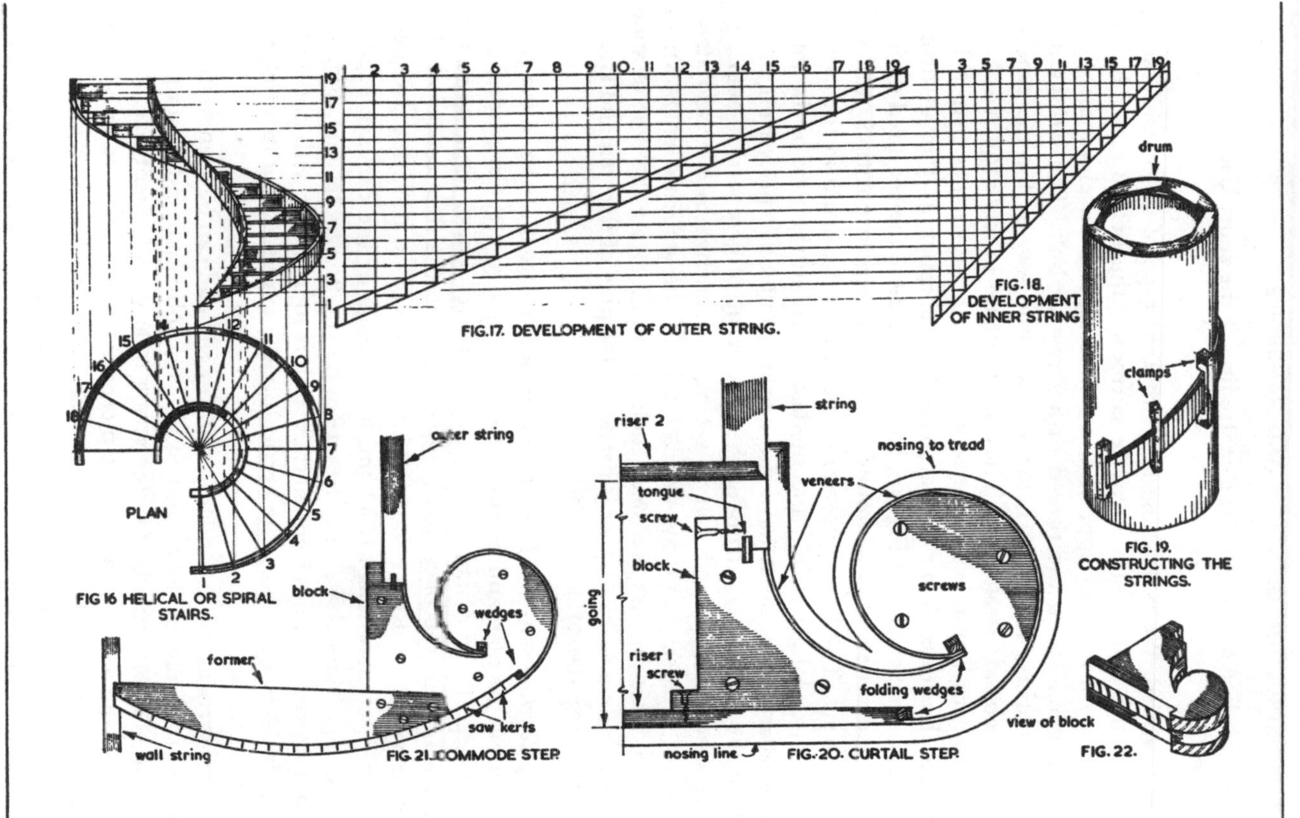

FIG 16 HELICAL OR SPIRAL STAIRS.

FIG.17. DEVELOPMENT OF OUTER STRING.

FIG.18. DEVELOPMENT OF INNER STRING

FIG.19. CONSTRUCTING THE STRINGS.

FIG.20. CURTAIL STEP

FIG.21. COMMODE STEP

FIG.22.

bisect this line to give e on the centre line of the stair. Use centre e for drawing the curve y–a–x.

Spiral stairs. We now come to a different type of geometrical stair, namely, the spiral stair or, more correctly, the helical stair. In the plan, Fig. 16, the stair appears to be circular, and the elevation shows that it is similar to the thread of a screw or bolt. The strings to a helical staircase are constructed in the same way as for the type of staircase already dealt with. They are built up to any required thickness by using a number of veneers. In a staircase such as that in Fig. 16, the thickness of the veneers could be as much as ¼ in. A drum, similar to that shown in Fig. 19, has to be made to the required dimensions, and veneers placed round it and clamped in position until the glue has set. The first veneer to be placed on the drum in the case of the outer or longer string, and the last veneer in the case of the inner or smaller string, must have the marking-out lines placed on them, as for the continuous string of the other type of geometrical stair.

It should be realized that to set out the strings to a helical stair, it is not necessary to draw the elevation of the staircase as shown in Fig. 16. This drawing has been included to illustrate the type of stair we are dealing with. All that is needed is the plan of the stair and knowledge of the rise of each step.

String development. To develop the outside or larger string to the stair, first draw the plan and number the risers 1, 2, 3, etc. On a horizontal line mark off the distances 1–2, 2–3, 3–4, etc. These dimensions can be taken from the plan. Remember that it is the first veneer of this string which must have the marking out placed on it, so the distances 1–2, 2–3, etc. are taken from the inside edge or the concave edge of the string. Mark these distances 1–19 on the horizontal line, Fig. 17, and drop vertical lines from all these points to intersect with horizontal lines brought over from the vertical line, giving the heights of all the risers to the stair.

This vertical line is similar to a storey rod and can be seen to the left of Fig. 17. Draw in the outline of the steps obtained and then mark in the top and bottom edges of the string to any convenient size. As can be seen, the string is a straight piece of timber, and has only to be placed round the drum at the correct rake or angle to produce the correct curvature. In the workshop the setting out of the string would be done on the surface of the drum and not on a large flat surface because this would probably not be practicable. The foregoing description is included to enable readers to understand the geometry involved.

The inner string is made in the same way. Another drum, to the required dimensions, is made, and the veneers to form the small string are cramped round it as for the larger string. As it is the last veneer which must have the marking out placed on its surface, the distances 1–2–3 etc. on the horizontal line have to be taken from the plan of the small string, Fig. 16, and on the edge nearest the outside or large string. This veneer will be clamped on to the drum last of all, and straight on to the top of the others already in position. Remember to place the veneers round the drums at the correct rake. It is quite simple to work out how high the top edge of the strings have to rise in passing round the drum the required distance. For instance, the strings in Fig. 16 have to travel upwards a distance of 19 risers in passing round the drum three-quarters of a turn. Remember, also, that the top edges of all the veneers have to lie in a horizontal plane when tested radially.

The steps to a helical stair can be of the traditional kind or can be of the open-riser type, which means that treads only are to be used. The strings can be recessed to take the ends of the treads or treads and risers in the usual way. Some modern staircases constructed in the helical style have a central string and the treads fixed to the string by means of wooden or purpose-made metal brackets. Sometimes these strings are as much as 18 in. to 2 ft. in depth. They are built up with ½ in. thick veneers, three or four being placed round the drum at a time, these being left for twenty-four hours when another four are added, and so on until the required thickness has been reached.

Handrailing. We now come to the problem of handrailing. As no newels are included in geometrical stairs, the handrails have to be continuous like the strings. To form the curved portion of handrail at the change of direction in the traditional way of handrailing, however, demands a greater knowledge of geometry, especially in handrailing involving two bevels.

Single-bevel work. The first two examples are single-bevel work ahd are not very difficult when compared with two-bevel work. When we consider all the other examples in this chapter, however, the reader will have to turn to Chapter 19 and learn how to develop oblique planes if he wishes to understand what is happening in this part of the work on handrailing. Let us consider the example in Fig. 23. At the bottom of the illustration is shown the plan of a staircase at the top of a straight flight of stairs leading to a landing. The handrail at this point is also shown. Above this drawing is the elevation of the last two steps in the flight which will, of course, give the pitch of the stairs. Notice that the curved portion of handrail in the plan is placed centrally over the quadrilateral a–b–c–d. This quadrilateral is the plan of a square prism with its top surface inclined at an angle, in this case equal to the pitch of the stairs (Fig. 26).

Preparing templets: To develop the shape of the templet required to make the curved part of the handrail, it is first necessary to develop the shape of the top surface of the prism, and from that develop the shape of the templet. Proceed as follows. Draw the plan of the handrail and the elevation of the top steps. Place in position the plan of the square prism so that sides b–c and c–d are in line with the centre lines of the handrail. Draw the pitch line of the stairs, and project up to the pitch line the edges a–d and b–c of the prism to give a and b on the pitch line. Project these points over at right angles and make a–d and b–c in the development equal to a–d and b–c in the plan. Then a–b–c–d in the development is the true shape of the top surface of the prism.

Divide the width of the plan of the prism into any number of parts and project these points up to the pitch line, and from here over to the c–d edge of the development. Make the distances a–a′, a–a″, 1–1′, 1–1″, etc. equal to those in the plan. Draw freehand curves through these points to obtain the shape of the templets required. Add two straight sections, say 2 in. long, on each end of the curved templet to complete its shape.

Two templets this shape are required for the wreath (this is what the curved portion of the handrail is called), and these can be cut from, say, ¼ in. plywood, see Fig. 25a. The thickness of the material required for the wreath is found by drawing a horizontal line to represent the top edge of the material (25c). Draw a line at any point at the same angle as the bevel shown in the elevation, Fig. 23, and from this angle construct a rectangle equal to the size of the handrail. Draw another horizontal line through the bottom edge of the handrail, and the distance between the two horizontal lines is the minimum thickness of timber required for the wreath.

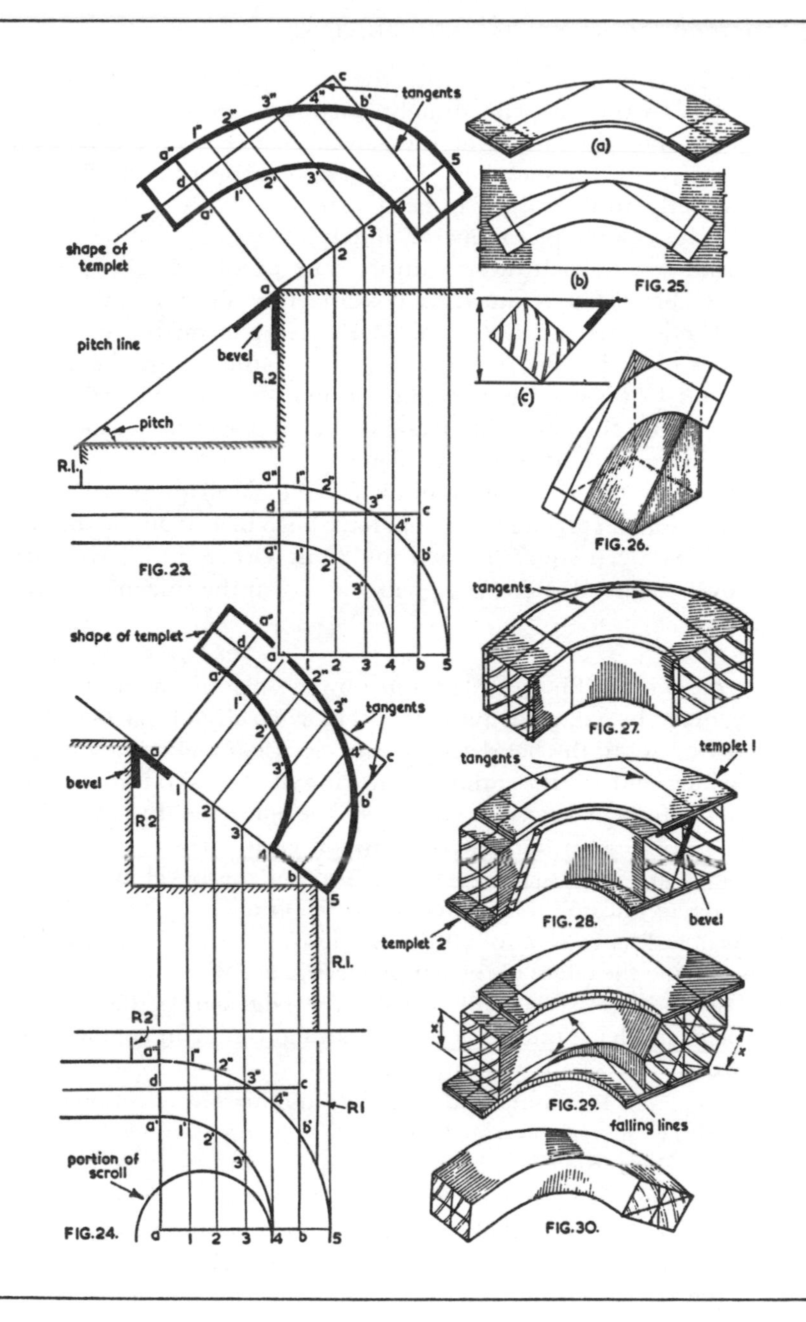
tangents
shape of templet
pitch line
bevel
R.2
pitch
R.1.
FIG. 23.
shape of templet
tangents
bevel
R2
R.I.
R1
portion of scroll
FIG. 24.
(a)
(b)
FIG. 25.
(c)
FIG. 26.
tangents
FIG. 27.
tangents
templet 1
FIG. 28.
bevel
templet 2
FIG. 29.
falling lines
FIG. 30.

Fig. 24 shows how the templet for the wreathed portion of a scroll is developed. In this case, no straight portion is required where the wreath joins the flat portion of the scroll. Notice that the tangent lines are placed on the templets. These assist in positioning the last named correctly on the timber to be shaped.

Fig. 25a shows a templet prepared and 25b how it should be placed on the plank from which the wreath is to be produced. If placed in this way it eliminates as much as possible the 'short grain' in the material. Having marked the shape of the templet on the plank its shape is cut out with the bow-saw or, better still, on the bandsaw, cutting about $\frac{1}{2}$ in. away from the lines around the curves, and about $\frac{1}{32}$ in. away from the ends. Afterwards, the ends can be cleaned up, perfectly square, with a smoothing plane.

Fig. 27 shows the wreathed portion for the scroll prepared ready for shaping. Note that the tangent lines have been placed on the timber, and these have been squared down the ends. Other lines have been squared over each end and exactly half way down the thickness of the material.

Marking the timber. The templets can now be placed on the timber, and care must be taken to position them correctly. First a sliding bevel must be set to the bevel seen in the elevation and a line marked on the wider end of the timber, allowing it to pass through the intersection of the lines squared across the end, see Fig. 28. The templets are placed on the material, one on each surface, so that the tangent lines are immediately over the ends of the bevel at one end and the tangent line on the wreath at the other end.

The next task is to remove all the timber on the wreathed portion which is outside the edges of the templets, Fig. 29. When this has been done it should be possible to place a straight-edge across the edges of the templets and have the surface of the straight-edge in contact with the shaped surface of the wreath.

The next step is to remove both templets and mark the section of the handrail on the ends of the wreath (Fig. 29). The top and bottom edges of the material are then prepared as in Fig. 30. This is done by carefully placing the falling lines on the wreathed portion in such a way as to avoid any irregularities and to assure a good sweeping curve to the handrail when completed, see Fig. 29. At all times the top and bottom edges of the wreath should be at right angles to the side surfaces.

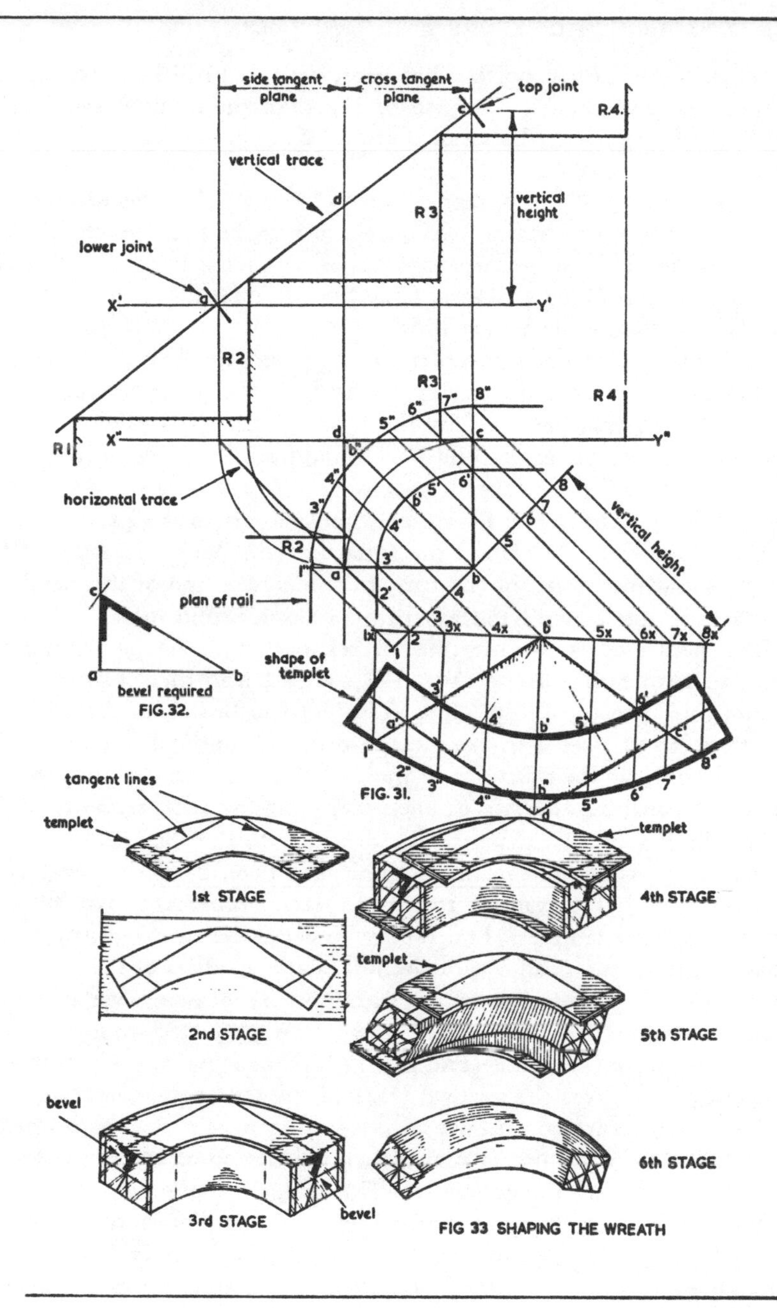
side tangent plane
cross tangent plane
top joint
R4.
vertical trace
d
c
vertical height
R3
lower joint
X'
a
Y'
R2
R3
7"
8"
R4
R1
X"
d
5"
6"
c
Y"
b"
horizontal trace
4"
5'
6'
8
vertical height
3"
b'
4'
7
6
5
R2
3'
1"
a
b
c
plan of rail
2'
4
3
3x
4x
b'
5x
6x
7x
8x
a
b
bevel required
FIG.32.
shape of templet
1x
2
1
3'
4'
b'
5'
6'
a'
c'
1"
2"
3"
4"
b"
5"
6"
7"
8"
d
FIG. 31.
tangent lines
templet
1st STAGE
templet
4th STAGE
templet
2nd STAGE
5th STAGE
bevel
6th STAGE
3rd STAGE
bevel
FIG 33 SHAPING THE WREATH

When the wreathed portion has been prepared in its square form it is moulded and then is ready to be fitted to the straight sections of handrails by using the handrail bolt and dowelled joints.

Two-bevel work. We now come to two-bevel work in handrailing in which, as the term implies, two bevels are applied to the wreath, one at each end. It is necessary to emphasize that a knowledge of the development of oblique planes, Chapter 19, is required to follow this part of the chapter on handrailing.

Let Fig. 31 be the plan of a turn in a staircase, this being two straight flights connected by a quarter-space landing, the space between risers 2 and 3 being that occupied by the landing. Since two-bevel work is based on a prism, as in one-bevel handrailing, it is necessary to superimpose the plan of the turn on to the plan of a prism. As the two flights are at right angles to one another, the prism is square.

Let a–b–c–d be the plan of the square prism. Make the sides of the prism equal the radius of the turn to the centre line of the handrail. The centre line of the handrail travels from a round to c.

The next step is to draw the stretch-out of the steps around the turn, so with centre d and radii d–R2 and d–a in turn, project these points round to the x′′–y′′ line. The vertical line from R2 will give the position of riser 2 in the stretch-out, and vertical lines from R3 and R4 will give the positions of these risers. The stretch-out can now be drawn, making the rise of each step suit the requirements of the stairs. In Fig. 31 the distance R2–d added to the distance d–R3 are equal to the width of one tread. This has been done deliberately in this first two-bevel example to simplify the problem, as will be seen later in other examples. The falling line can be drawn through the nosings giving the positions of the joints at the edges of the side- and cross-tangent planes. These are obtained by drawing vertical lines from c, d and the point where the curve projected round from a intersects the x′′–y′′ line. The falling line, as it passes over the side-tangent plane, gives the vertical trace of the top surface of the prism, this being the surface we have to develop to obtain the shapes of the templets. The x′–y′ line is drawn through the point where the falling line intersects with the edge of the side tangent plane.

Reference should now be made to the oblique planes in Chapter 19 to see how the surface a–b–c–d is developed, Fig. 58. Having obtained this shape, a′–b′–c′–d′, one can proceed to develop the shape of

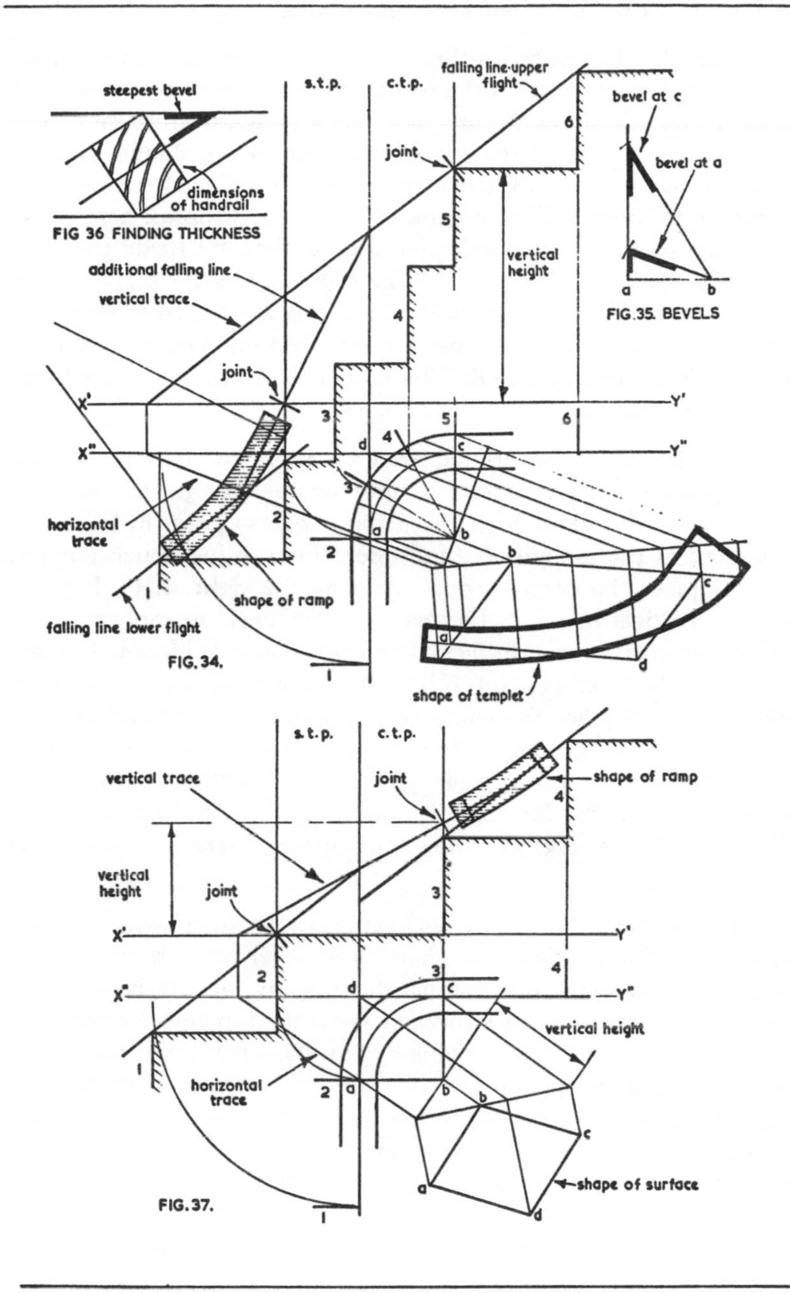
steepest bevel
dimensions of handrail
FIG 36 FINDING THICKNESS
s.t.p.
c.t.p.
falling line-upper flight
bevel at c
bevel at a
joint
vertical height
additional falling line
vertical trace
FIG.35. BEVELS
joint
X'
Y'
X"
Y"
horizontal trace
shape of ramp
falling line lower flight
FIG.34.
shape of templet
s.t.p.
c.t.p.
vertical trace
joint
shape of ramp
vertical height
joint
horizontal trace
vertical height
shape of surface
FIG.37.

the templets. To do this the outline of the curved portion of handrail must be placed on the plan of the turn. The distance from 1–8 should be divided into a number of parts, and lines drawn from these points across the plan parallel to the horizontal trace. They should then be projected down to the 1x–8x line and from this line downwards at right angles to 1x–8x. The lengths of the lines running across the development of the surface a–b–c–d should then be made to equal those across the plan of the turn to give points on the shape of the templets. Straight sections about 2–2½ in. long are added to each end of the templet development, these straight sections being parallel to the tangent lines a′–d and c′–d. The ends of the templet must be at right angles to the tangents a′–d and c′–d.

Bevels. To obtain the bevels required to apply to the ends of the wreathed portion, first construct a right angle, Fig. 32, making a–b equal to a–b in the plan. Next place the compass point in b′ in the development of the top surface, and open them to just touch tangent a′–d. Next place the compass point in b on the right angle, Fig. 32, and cut the vertical arm in c. This will give the bevel to apply to lower end of the wreath, near a, in fact. The other bevel is obtained in the same way. With compass point in b′ open the compasses to just touch tangent c′–d, and place the compass point in b Fig. 32 and cut the vertical arm of the right angle. It will be found in this case, that the vertical arm will be cut in c, making the bevel for the top end of the wreath the same as that for the lower end. This is not always the case, as will be seen later. Fig. 33 shows the various stages in shaping the wreath.

Placing the templets. The only difference in these drawings is the positioning of the templets. As there is a bevel at each end of the wreath, care must be taken to see that the tangent lines on the templets are immediately over the lines running from the top and bottom ends of the bevels, and parallel to the tangents on the wreath (see 4th stage).

Fig. 34 is the development of the shape of the templets for a staircase of two straight flights connected by a quarter-space of winders. The stretch-out of the steps and the development of the surface a–b–c–d is straightforward if note has been made of the last example and the geometry of oblique planes.

There is, however, a complication in this and similar examples. The falling lines of the lower and upper flights do not meet, and as it is necessary for them to meet to obtain a continuous handrail round the turn, it is essential to introduce another falling line so that this will connect the other two. This also means introducing a ramped portion of handrail in between the wreath and the lower straight handrail, as seen in Fig. 34.

The top falling line should be brought down to the vertical line immediately above d in the plan, and from here the additional falling line placed in to meet the lower falling line near riser 1. This will allow the ramp to be included—its end to be some distance, say 2 in., away from the lower joint line of the wreath, which, incidentally, must fall on line x′–y′ and the edge of the side-tangent plane. The top joint, of course, is on the edge of the cross-tangent plane.

The vertical trace is the continuance of the top falling line, and should extend down to the x′–y′ line, which passes through the lower joint, from here vertically down to the x′′–y′′ line to obtain the position of the horizontal trace.

Two bevels are required, as seen in Fig. 35, and these are developed in the way already explained. The thickness of material is found by taking the steeper of the two bevels and proceeding as before, see Fig. 36.

Fig. 37 is another staircase of two flights connected, this time by a quarter-space landing. The prism plan a–b–c–d can be seen, and the stretch-out of the steps shows that the two falling lines, again, do not meet. This time, however, the lower falling line is above the position of the upper falling line, and so it is necessary to introduce a ramp at the upper end of the curved portion of handrail. Having ascertained the position and shape of the ramped portion and the development of the templets, the bevels and thickness of material required can proceed as before.

The next problem, involving a turn in a staircase which is not a right angle (see Fig. 38), also necessitates the introduction of a ramp. The prism on which this problem is based is trapezoidal in plan. The staircase consists of two straight flights connected by winders with the risers numbered 2, 3 and 4. Again, the falling lines do not meet and a ramp must be used, this time at the lower end of the curved portion. A study of this type of oblique surface, Fig. 60, Chapter 19, will assist the reader in understanding the method used for developing the surface a–b–c–d and also the shape of the templets. The bevels are shown in Fig. 39, and are developed as before.

Some readers may find it difficult to understand the geometry used for developing the oblique surfaces in the problems on handrailing. Fig. 40 has been included in this chapter for guidance. First try to imagine that the prism (a square one in this case) has been placed in the angle set up by the vertical and horizontal planes, Fig. 40. a–b–c–d is the top surface which has to be developed. The edge c–d is projected down to the x–y line and this projected line is the vertical trace of the top surface. Another line from where the vertical line meets the x–y line and passing through the lowest corner of the top surface will give the horizontal trace. We must now try to imagine a right-angled triangle, a′–c–c″, standing at right angles to the horizontal trace and in contact with the corner of the prism of which b is the top point. The edge c–c″ is equal to the vertical height between the lowest and highest corners of the prism—a and c.

It should also be noted that all the lines which start from the top edge of the triangle and go across the top surface of the prism are parallel to the horizontal trace, and are therefore all horizontal lines and can be measured in the plan.

If the triangle a′–c–c″ were turned so that it lay on the horizontal plane as seen in Fig. 40 and the various lines projected downwards at right angles to the line a′–c′ (which is now the top edge of the triangle), and all those lines made equal in length to those across the top surface of the prism, not only can the shape of the top surface be drawn but the curve a–c can also be plotted.

Dancing steps. Before leaving the subject of staircase work it will be well to consider a stair with dancing steps, Fig. 41. This staircase must not be confused with those in publications illustrating the new Building Regulations, where certain requirements are now in existence regarding winders in staircases. Readers must refer to these to ascertain what is actually required. To overcome the danger of very narrow steps near the newel post at the change of direction, it is possible to increase the widths of the winders by introducing additional tapered steps (dancing steps). From the plan Fig. 41 it will be seen that there are six of these instead of the usual three at the turn. The walking line of a staircase is the path of an individual ascending or descending the stair and this is considered to be approximately 16 in. from the outer string. When possible, it is wise to have the going of all the steps of a stair equal at the walking line, and so this should be established before

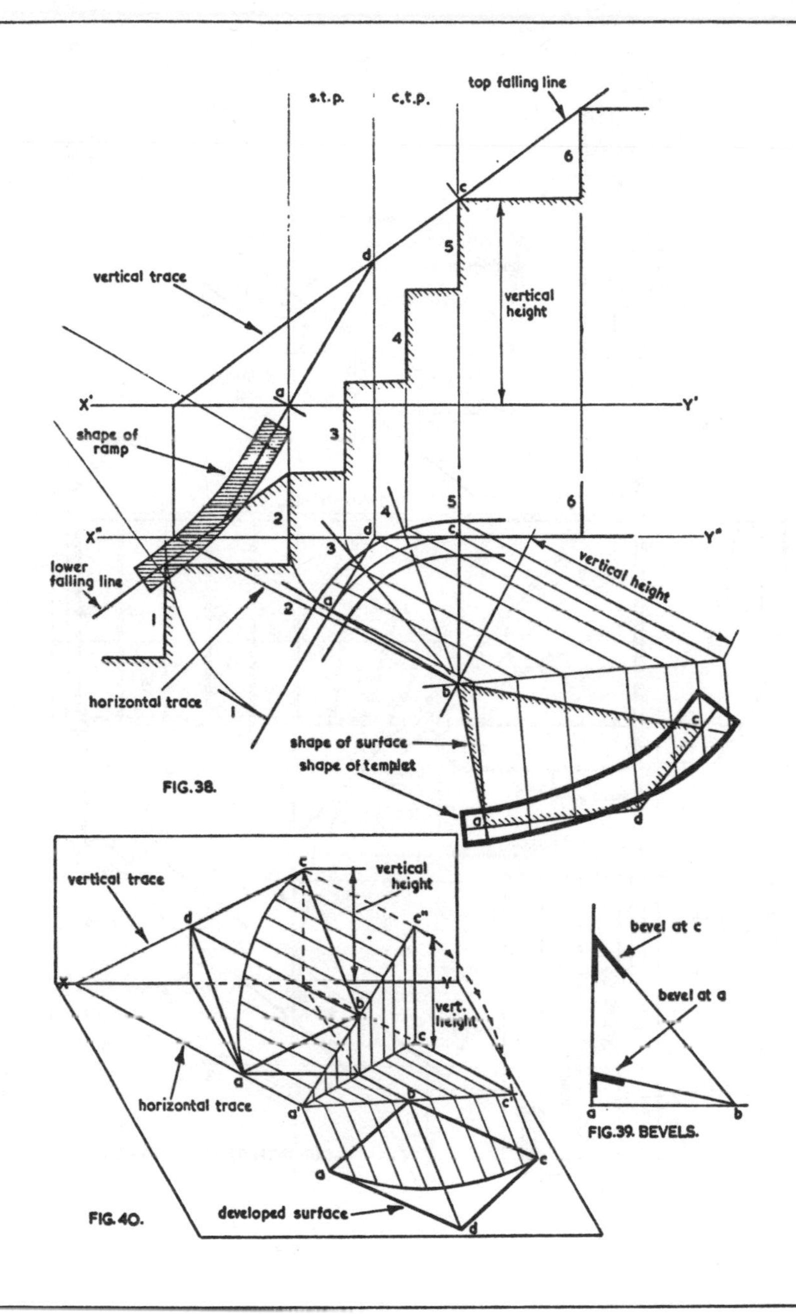

FIG. 38.

FIG. 39. BEVELS.

FIG. 40.

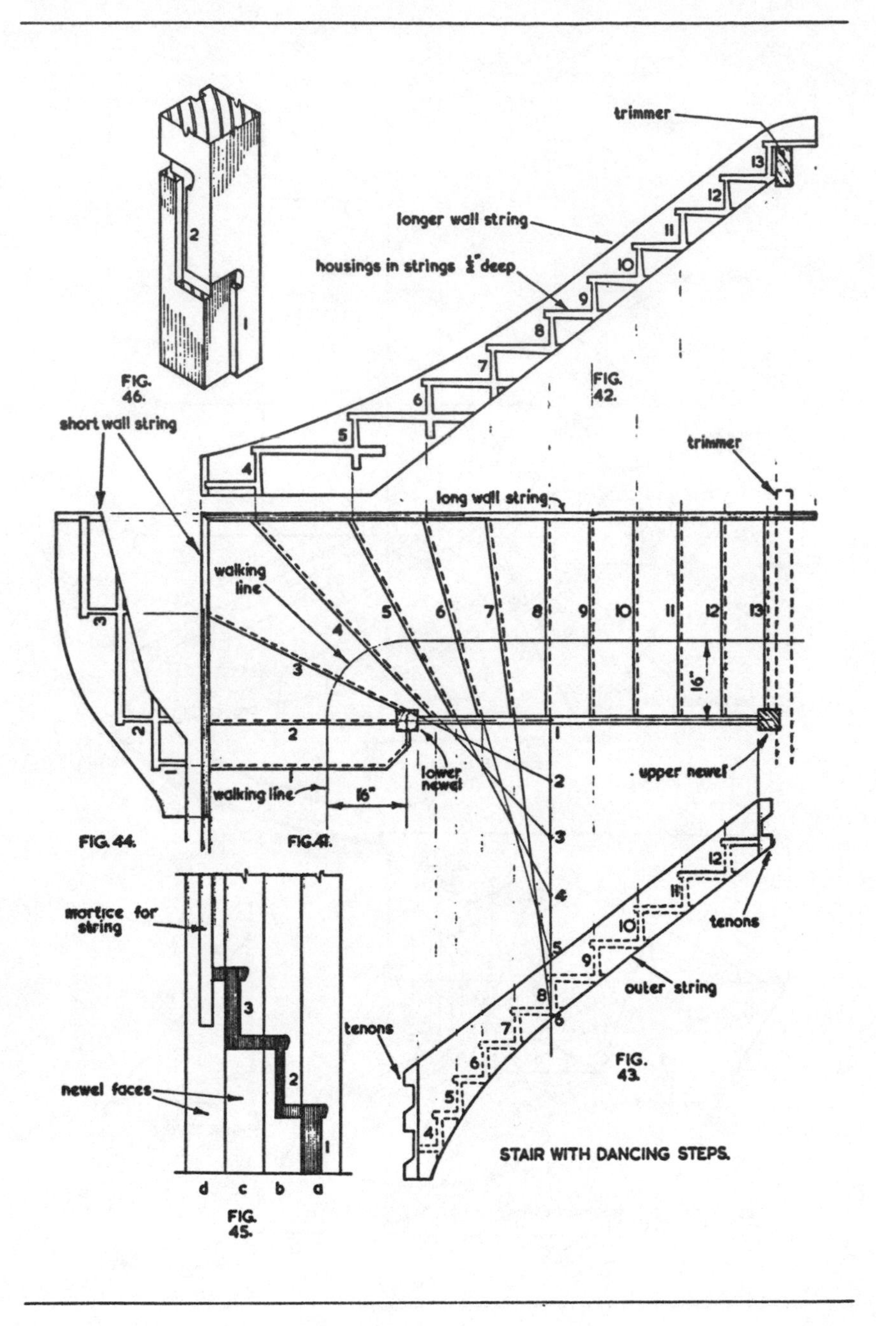

STAIR WITH DANCING STEPS.

setting out the positions of the risers. Having drawn the walking line on the plan and marked the 'going' on the walking line, it is necessary to decide on the number of dancing steps one is prepared to include in the flight.

To place on the drawing the positions of the risers to the dancing steps, draw a vertical line down from the first riser which is at right angles to the strings, in this case riser number 8, and mark off on this line from the centre of the outer string a number of spaces, say 12 in. apart. The number of spaces should be equal to the number of dancing or tapered steps. From these points lines can be drawn through the points on the walking line to obtain the positions of the risers to the tapered steps. It may be necessary to adjust the distances 1–2–3 etc. until a satisfactory positioning of the winder risers has been obtained.

Also shown on the drawing are methods for obtaining the shapes of the strings and the positions of the step housings, and it will be seen that the outer string has to be increased in width considerably towards its lower end because the falling line is much steeper at this point. This is due to the fact that the width of each step from number 8 downwards is narrower at the outer string than it is where it enters the long wall string. Fig. 45 shows how the four faces of the lower newel are set out for recessing, and Fig. 46 is a pictorial view of the newel. Although this type of staircase produces one which is safe and easy going, one must also remember that it is more expensive to produce.

18 Glulam Work

The word 'Glulam' is a term used to describe a method of building up a timber component of any shape to any requested dimensions, and involves gluing together a number of laminations on a specially prepared cramping area. Although this has become a highly specialized section of the woodworking industry in recent years, a brief description of the procedure used in producing a glulam item will help the woodworker should he be called upon to assist in this type of production.

Small work. For small pieces of curved work made in a similar manner to that to be described in these pages, no special preparation is necessary. Fig. 1 shows how small items of work can be manufactured. There is a flat base board, say of 1 in. blockboard or plywood, strengthened if necessary by battens glued and screwed to the underside. The double outline of the item to be produced is marked carefully on the surface, and formers, cut to the shape are screwed on to coincide with the inside of the two lines.

Glue is spread on both surfaces of the laminations except the two outside pieces which should have only one surface glued. The pieces are loosely placed together on to the base board, and a second set of formers, cut to coincide with the outer outline of the item to be produced, are placed behind the laminations. By using a number of G cramps across the sets of formers the work can be cramped up to bring the surfaces of all the laminations into close contact, see Fig. 1.

Large work. For structural laminated work much greater care must be taken. For instance, the selection of the right grade of timber is important. There are two groups of softwood timbers which are recommended for structural purposes, the first including Douglas fir, longleaf pitch pine, and shortleaf pitch pine. The second group

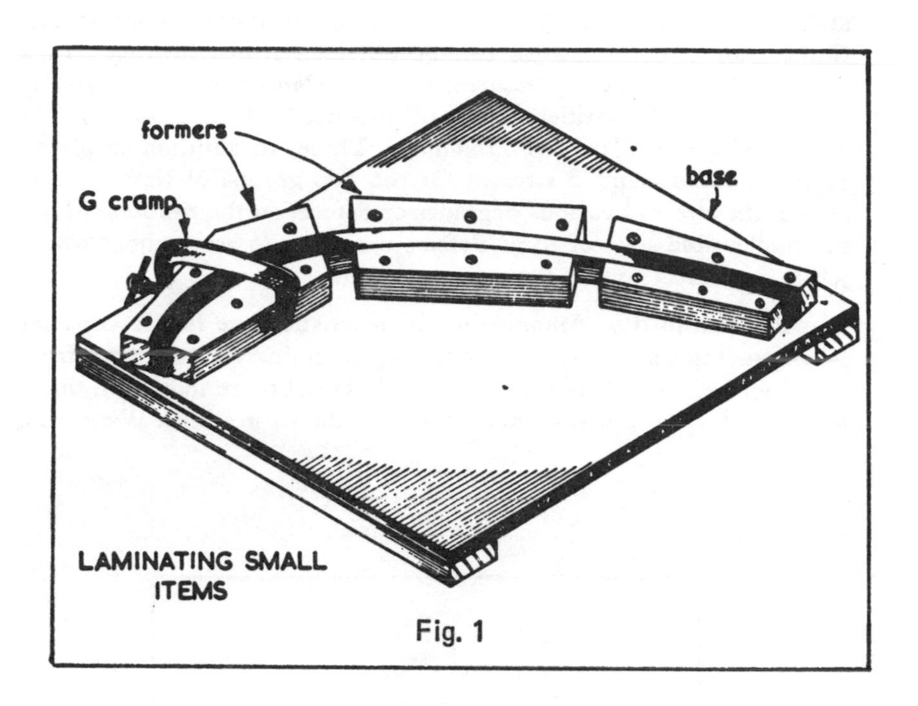

LAMINATING SMALL ITEMS

Fig. 1

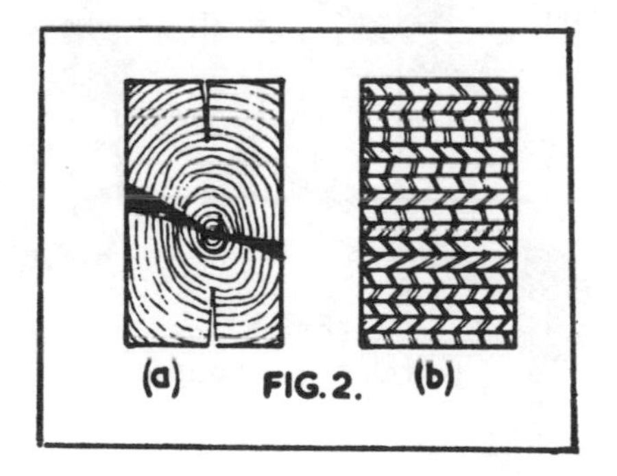

FIG. 2.

includes Canadian spruce, European larch, redwood, whitewood, and western hemlock. More information on these recommended timbers for structural work can be obtained from B.S.1860: 1952 *Structural Softwood—Measurement of Characteristics Affecting Strength,* and the British Code of Practice C.P. 112 (1952) *The Structural Use of Timber in Buildings.* These, in addition to gluing tables of recommended stresses for the two groups of timbers, also set out the effect of various degrades or defects on the timbers. Maximum allowable sizes of these defects are given to assist those whose job it is to select timber for structural work.

Timber suitability. Among the characteristics one has to consider when grading softwoods is the slope of the grain. This can vary from 1 in 8 for beams and compression members not more than 4 in. thick, to 1 in 11 for compression members more than 4 in. thick. Wane, too,

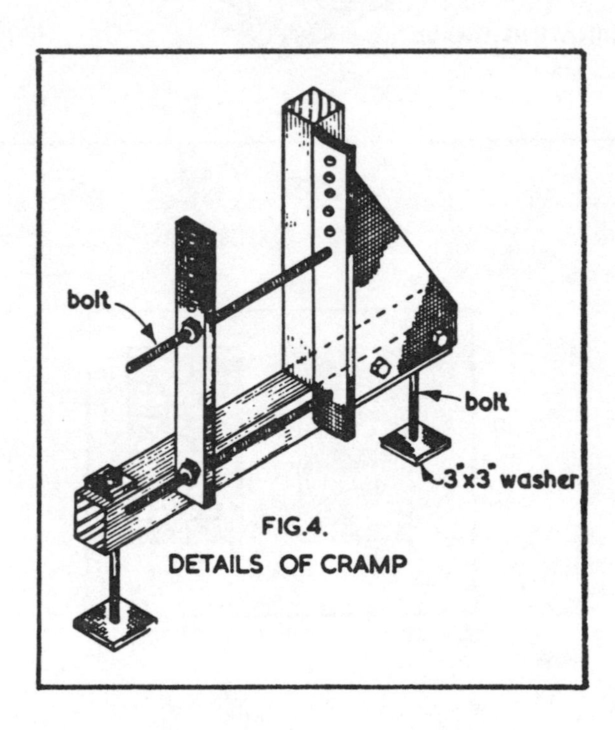

FIG.4.
DETAILS OF CRAMP

must be considered and so must the rate of growth, because softwoods derive their strength from the tracheids of the latewood or summer growth. A slow rate of growth usually indicates a strong timber. Knots, shakes and splits affect the strength of a piece of timber and so these have to be measured and taken into calculation.

One advantage of using timber for glulam work is that, since it is formed from laminations, defects such as knots can be dispersed more evenly throughout the finished product. Fig. 2 shows the difference between a solid beam a, and a laminated beam, b. The solid piece can be much weaker than the laminated one because serious defects can be concentrated at one or more points.

Many firms now producing glulam components have their own methods for producing such work, and most install what they consider to be the best kinds of cramps for the work they are producing. Fundamentally, however, all systems are much the same. Many of the cramps or chairs, as they are often called, are made from metal angles and channels but basically they are similar to that shown in Fig. 4. This consists of two pieces of timber jointed to form a right angle with a mild steel bracket bolted to them. The bracket should be made from mild steel plate $\frac{3}{8}$ in. thick, with one of its edges turned at right angles to the main portion which is bolted to the timbers. In this returned edge are drilled a series of holes through which bolts

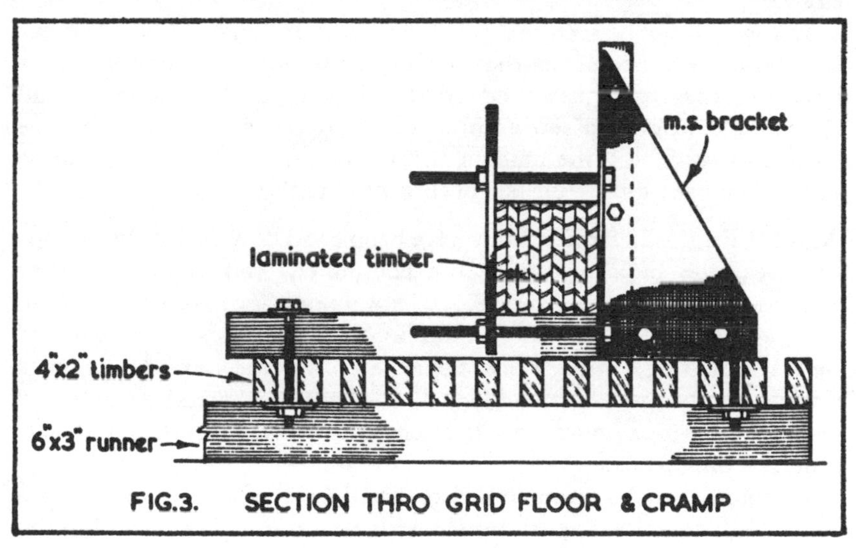

FIG.3. SECTION THRO GRID FLOOR & CRAMP

pass. The other edge of the cramp consists of a piece of $\frac{3}{8}$ in. mild steel plate with bolt holes drilled through it to coincide with those in the bracket. Two more bolts with large square washers are provided to enable the cramp or chair to be fixed in position on the cramping area.

Cramping area. This should be made so that cramps can be fixed down on to the surface in any desired position. One method consists of a floor of slats fixed to runners such as that in Fig. 3. Runners 6 in. by 3 in. are secured to the concrete or timber floor, and 4 in. by 2 in. or 4 in. by 3 in. timbers fixed to the runners, with spaces between, to form the floor of the cramping area. The drawing also shows a cramp fixed to the slatted floor by means of the two bolts passing through the slats with large square washers below.

Fig. 6 shows the plan of a number of these cramps placed to produce a glulam component in the shape of a parabolic curve. To produce this it is necessary to set out the curve, full size, in the workshop so that the cramps can be positioned correctly. If it were considered necessary to produce a templet to assist in the cramp positioning, one method for producing it is seen in Fig. 5. This consists of drawing lateral lines at right angles to a datum line at regular intervals, and scaling the lengths of the laterals from the architect's or designer's drawing.

In Fig. 5 it can be seen that the datum is 19 ft. long and perpendicular laterals have been constructed every 12 in. along its length. These have been carefully measured on the drawing and transferred, full size, to the workshop setting-out. The templet, when made, can be used for positioning the cramps and for trimming the component to shape when the gluing up has been completed.

Assembling. When the cramps have been fixed down on to the cramping area, the building up of the component can commence. For an item such as that shown in Fig. 6 laminations approximately $\frac{3}{8}$ in. to $\frac{1}{2}$ in., according to the shape, would be used. Up to six of these would be assembled at a time, these being left until the next day when a similar number would be added. This procedure would be repeated every twenty-four hours until the component had been built up to its required dimensions.

It would then be manhandled or lifted by mechanical means to a convenient position for trimming with a portable saw and cleaning up to size and finish with a portable electric plane and sander. Lastly it would be varnished or sealed, whichever was appropriate, and any additional work carried out before delivery to the site.

Figs. 9 and 10 show an item before and after finishing. Fig. 10 illustrates how the part would appear after removal from the cramps, and Fig. 9 how it would look when completed. It should be obvious from these drawings that a templet is necessary to bring the item to the required shape.

Figs. 7 and 8 show two methods of fixing for glulam work. Fig. 7 has a mild steel strap for holding two halves of an arch at the crown, and Fig. 8 a mild steel base bracket for fixing one of its feet.

Surfaces to be glued should be planned to fairly accurate dimensions. Glue can be spread by a brush on a small job, but for the larger type of work, such as roof arches and the like, a mechanical glue spreader is desirable.

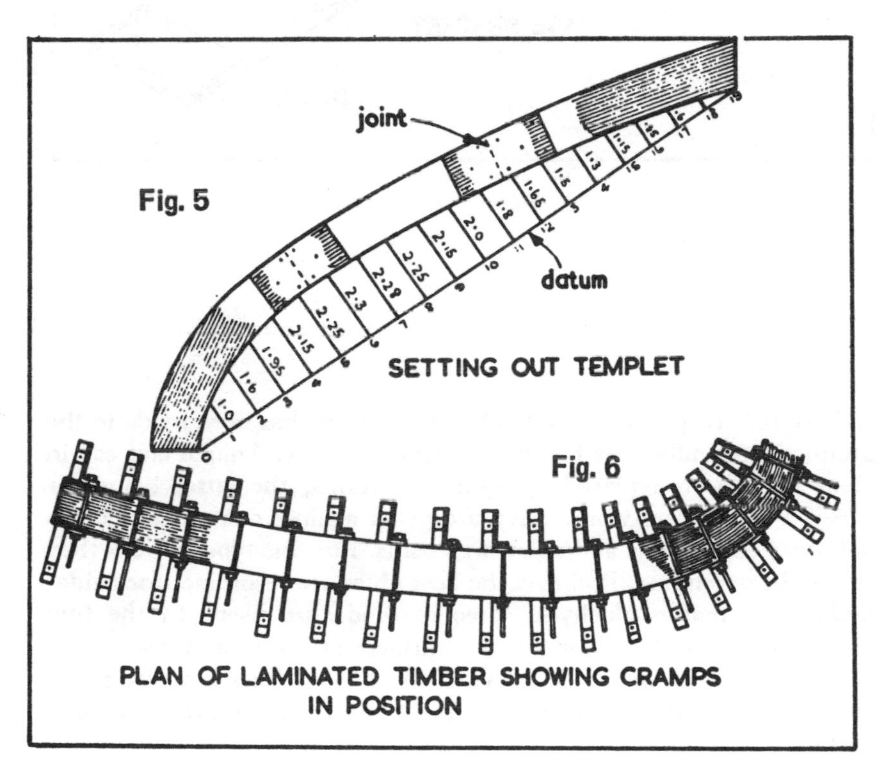

SETTING OUT TEMPLET

PLAN OF LAMINATED TIMBER SHOWING CRAMPS IN POSITION

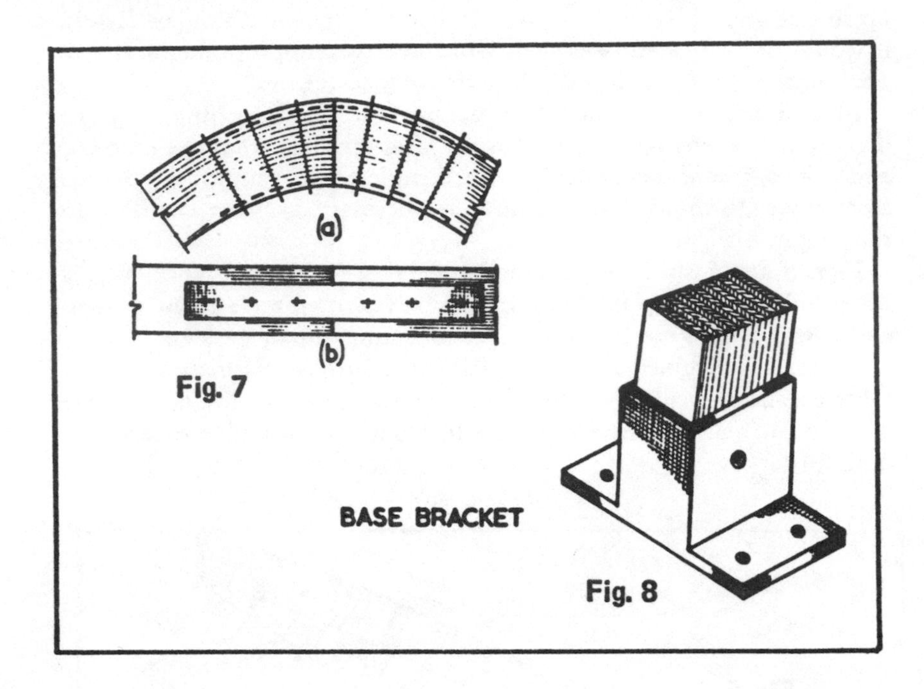

Fig. 7

Fig. 8

Adhesives. In recent years much advancement has been made in the production of adhesives for the building industry. Animal and casein glues have been replaced by synthetic resins, the latter being far superior for glulam work. They are proof against dampness and are not attacked by fungi and other organisms. For this type of work they should be gap filling. Probably the best glues are resorcinol-formaldehyde and ureaformaldehyde. Two methods are used. In the first the resin is spread over one of the surfaces and the hardener on the other, the setting commencing when the two surfaces are brought in contact. In the second method the resin and hardener are mixed and spread in one application.

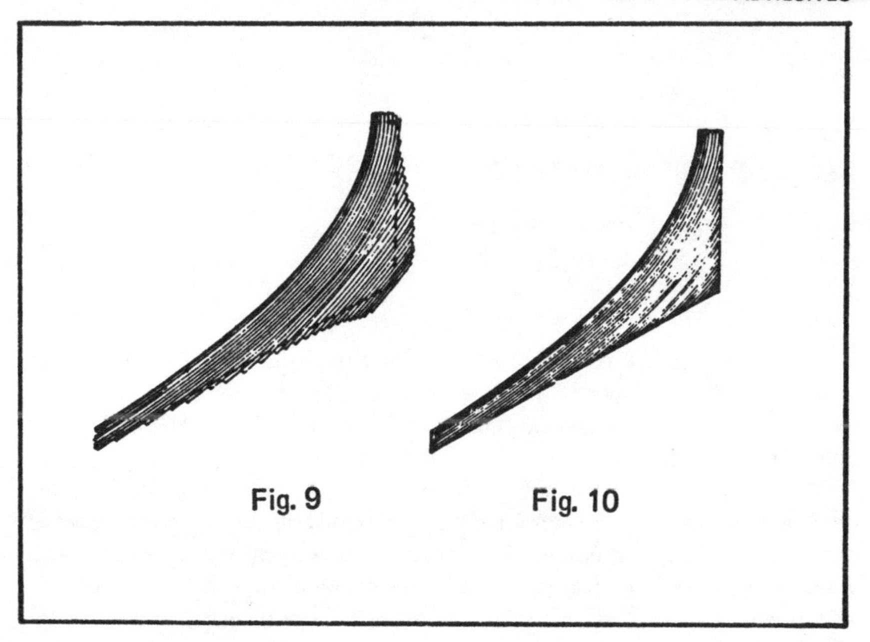

Fig. 9 Fig. 10

19 Applied Geometry

A large percentage of the work of the carpenter and joiner, in the advanced stages at least, involves geometry, and it is the purpose of this chapter to give the craftsman an idea of how a knowledge of geometry can be applied to practical work.

When straight and curved lines join, as in the case of mouldings and, as Fig. 7 shows, in the construction of templets, etc., it is necessary to ensure that no irregularities occur at these points. Therefore the straight line which joins a curved line must always be tangential to the curve. The first six illustrations explain how internal and external tangents to curves can be constructed, and Fig. 7 shows how this knowledge can be applied in a practical manner.

Tangents. Fig. 1 gives the method of drawing a tangent to a circle from any point p outside the circle. Draw the circle and mark point p in the desired position. Join p to the centre of the circle, a, and bisect p–a in b. With compass point in b and radius b–p describe a semicircle to cut the circle in c. Point c is the point of contact between the circle and the tangent.

Fig. 2 shows how to draw an external tangent to two unequal and touching circles. Draw the circles, any diameter, so that they touch in d. Join the centres a and b with a straight line and bisect a–b in c. With centre c and radius c–a draw the semicircle a–e–b. From d draw d–e at right angles to a–b to give point e on the semicircle. With centre e and radius e–d describe the semicircle f–d–g. Points g and f are the points of contact for the tangent g–e–f. g–a and f–b are normals and are at right angles to the tangent.

Fig. 3 shows how to draw an external tangent to two unequal circles which are some distance apart. Draw the circles and a line to join their centres a and b. Bisect a–b to give c, and with centre c and radius

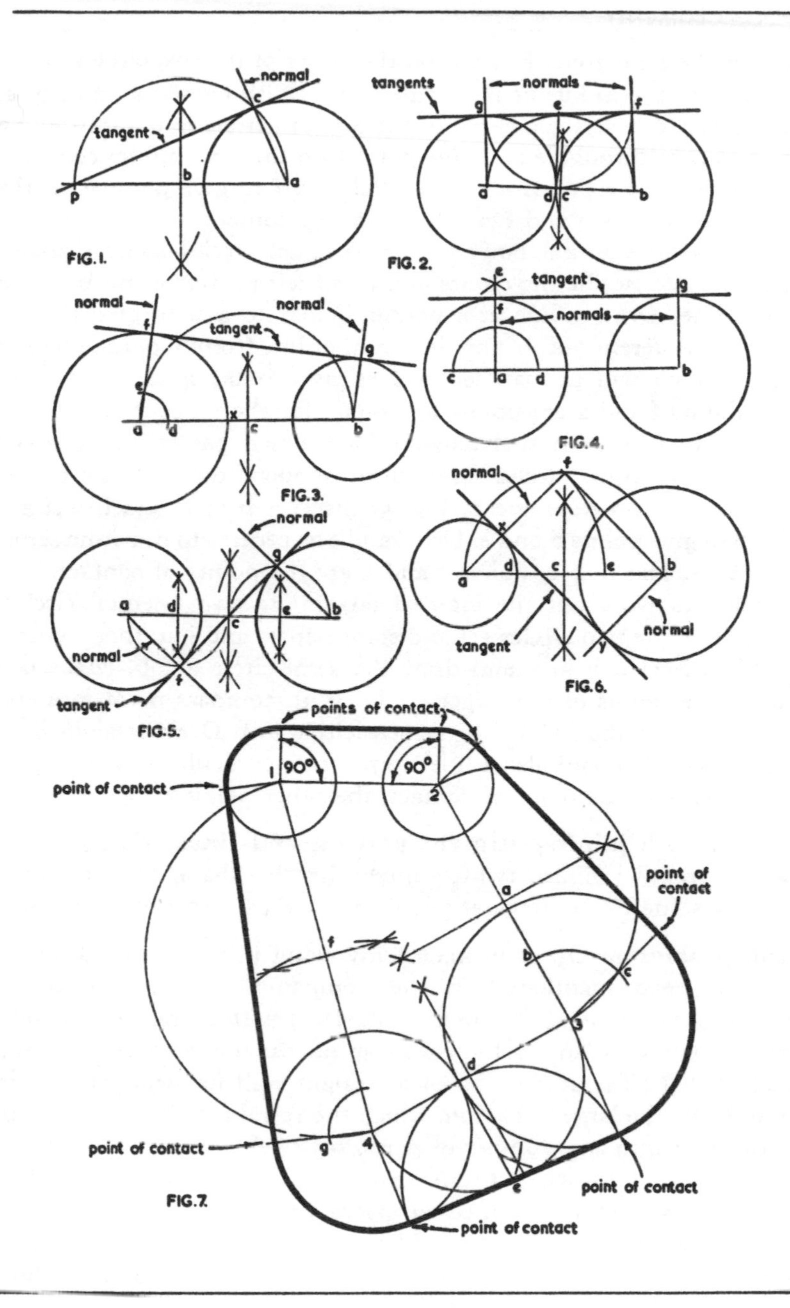
normal
tangent
FIG. 1.
tangents
normals
FIG. 2.
normal
tangent
normal
FIG. 3.
tangent
normals
FIG. 4.
normal
normal
tangent
FIG. 5.
normal
tangent
normal
FIG. 6.
points of contact
point of contact
90°
90°
point of contact
point of contact
point of contact
point of contact
point of contact
FIG. 7.

c–a draw the semicircle. From x, on the larger of the two circles, make x–d equal to the radius of the small circle. With compass point in a, and radius a–d describe a curve to cut the semicircle in e. Draw a line from a to pass through e and give point f on the circumference of the large circle. Add a line from b, parallel to a–f to give point g on the small circle. Points g and f are the points of contact.

To draw an external tangent to two equal circles some distance apart, Fig. 4, describe the circles and join their centres a and b with a straight line. Draw a line at a perpendicular to a–b to give point f on the circumference of that circle. Another line from b, perpendicular to a–b (which will be parallel to a–f) gives point g on the second circle. Points f and g are points of contact for the tangent.

To construct an internal tangent (one which passes between the circles) to two equal circles some distance apart, describe the circles and join their centres a and b, Fig. 5. Bisect a–b in c, and bisect a–c and c–b to give points d and e. Use d and e as centres to draw the semicircles a–f–c and b–g–c. Points f and g are the points of contact.

Finally, to work out the internal tangent to two unequal circles, Fig. 6, draw the two circles some distance apart and join their centres a and b. Bisect a–b in c and draw the semi-circle a–f–b. Make d–e equal to the radius of the larger circle. With compass point in a and radius a–e draw the arc to cut the semicircle in f. Draw straight lines from f to a and b, and also a line from b perpendicular to f–b to give point y. This is one point of contact, the other being x.

Templet with joining straight and curved lines. Fig. 7 shows how an irregular figure (which might be the shape of a templet) involving some of the tangent problems in Figs. 1–6 can be tackled.

Scroll problems. Other problems involving parts of circles which have not been mentioned in the companion volume *(Practical Carpentry and Joinery)*, are those concerned with spirals and scrolls. Fig. 8 shows how an Archimedean spiral can be set out. It could be adapted for use at the end of a straight wall handrail as seen in Fig. 9. Draw the large circle into which the spiral is to be constructed, and divide it into any number of parts, say twelve, and number these as shown on the drawing. From point 1 and along the diameter line mark off the same number of equal spaces, making the overall distance 1–13 the depth of the handrail. Using the centre of the circle and opening the compass to the various points on the diameter line,

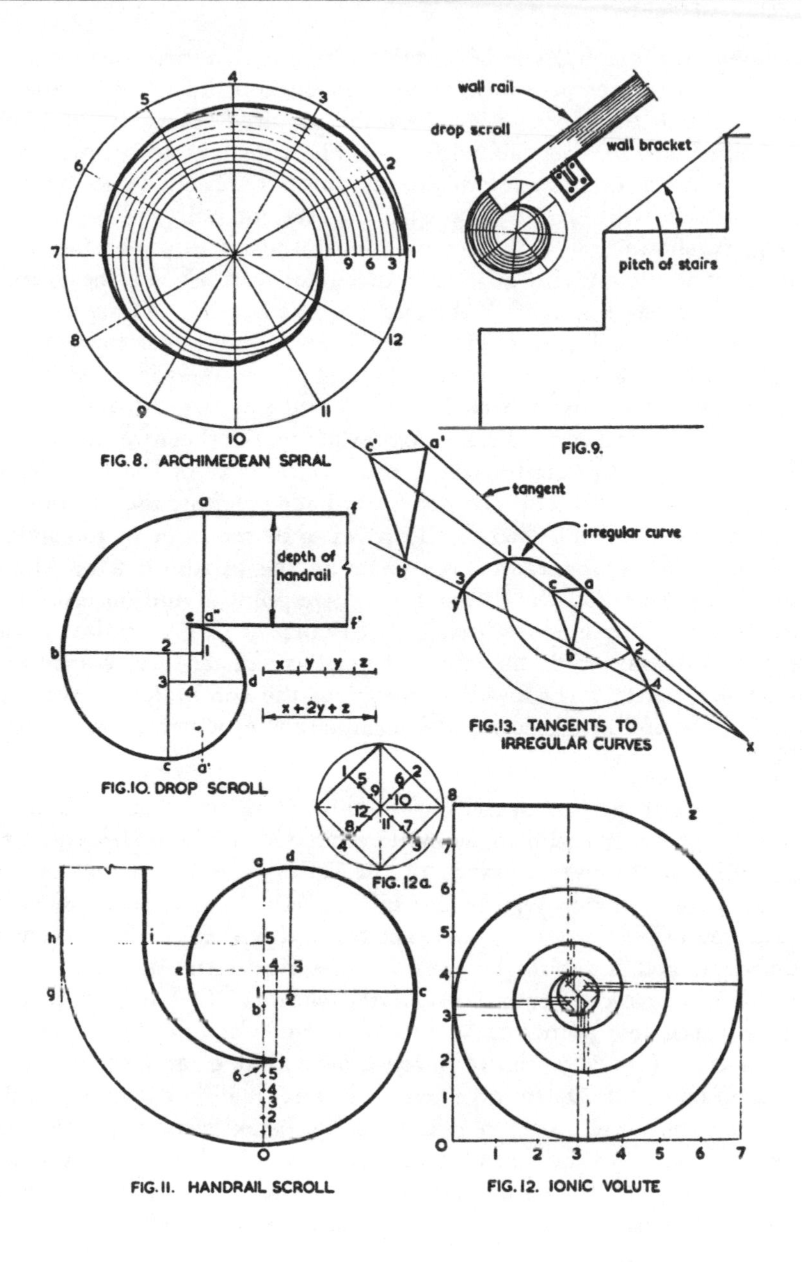

FIG. 8. ARCHIMEDEAN SPIRAL

FIG. 9.

FIG. 10. DROP SCROLL

FIG. 13. TANGENTS TO IRREGULAR CURVES

FIG. 12a.

FIG. 11. HANDRAIL SCROLL

FIG. 12. IONIC VOLUTE

positions on the spiral can be obtained by swinging the various arcs round to the appropriate normal, as seen in the drawing. A freehand curve through these points will give the required spiral. If this type of scroll is used for the wall handrail in Fig. 9 a slight adjustment will have to be made on the curve from, say, normal number 11 so that the scroll portion will not interfere with the straight rail.

Fig. 10 shows how a drop scroll can be set out to a pleasing line. This is constructed with quadrants of circles. Let a–a′ be the overall depth of the scroll, and a–a′′ the depth of the handrail. Draw the two horizontal lines a–f and a′′–f′. Next, draw a line equal to a–a″ in length, and divide it up into sections equal to $x + 2y + z$, as seen in the drawing, ensuring that x is larger than y and y is larger than z.

On line a–a′ and some distance below a′′ mark off centre number 1. The position of this centre will determine the size of the completed scroll, and it may therefore be necessary later to adjust the position of this centre if the completed scroll proves to be too large or too small.

With centre 1 and radius 1–a describe the quadrant a–b. Along the line 1–b mark off the distance x to give point 2, and describe the quadrant b–c; on line 2–c mark off distance y to give point 3 and describe quadrant c–d; on line 3–d mark off distance z to give the remaining centre to be used to complete the scroll. If the work is accurate the end of the scroll will terminate at e, which is on the line a′′–f′.

Scroll to geometrical stair. Fig. 11 shows how to set out a handrail scroll at the lower end of a geometrical staircase. Make O–a equal to four-fifths of the overall width of the scroll, and O–f equal to the width of the handrail. Divide O–f into six equal parts. Bisect O–a in b and step off towards a, one-sixth of the width of O–f. This is centre number 1, and it should be used for describing the quadrant O–c. On line 1–c mark off two-sixths of the width of O–f to give centre 2 and, using this point, draw quadrant c–d. On line 2–d mark off $1\frac{1}{2}$ sixths O–f to give centre 3 for quadrant d–e, and on line 3–e mark off one-sixth O–f to give centre 4. The quadrant from e should terminate opposite point 6 and directly below centre number 4. Make c–g equal to the width of the scroll, and with radius 1–g place the compass point in O and mark off the centre 5 on O–a. With centre 5 describe quadrants O–h and 6–i to complete the scroll.

Ionic volute. Fig. 12 is an Ionic volute, and fits into a rectangle eight units high and seven units wide. Its centre eye is one unit in diameter and is placed between points 3 and 4 on the vertical arm with its centre immediately over point number 3 on the horizontal arm. Fig. 12a shows how the centre eye is set out and numbered, each being used in numerical order to construct a quadrant. Accuracy is most important in this as in other scrolls if a pleasing result is required.

Irregular curve tangent. Fig. 13 shows how to draw a tangent to an irregular curve. Draw the curve y–z any shape, and assume that a tangent is required at point a. With centre a draw two arcs to give points 1 and 2, 3 and 4 on the curve. Draw a straight line through points 1 and 2 and another through points 3 and 4 to meet the first line in x. From x draw the tangent through a.

If point x is inaccessible, any triangle is constructed at a to give points b and c on the lines passing through the curve, and then another triangle, similar to the first (all sides parallel to those of the first triangle) to obtain point a′. Draw a line from a′ through a to obtain the tangent.

Solid geometry. Let us now turn our attention to solid geometry. The study of geometrical solids such as prisms, cylinders, pyramids, etc., is essential to the carpenter and joiner because most of the shapes found in the woodworking industry can be related to one or more of these solids. The basic developments of the common geometrical solids were dealt with in the companion volume, *Practical Carpentry and Joinery*. Here it is the aim to show how these developments can be applied to practical work.

Rectangular sections. Let us take rectangular prisms, or rectangular pieces of timber first. A simple case could be a square post supported by two square struts set at an angle of 60° (Fig. 14). The strut to the left shows that it has one of its surfaces facing directly upwards, and that to the right has one of its corners pointing upwards. The problem here is to produce the bevels so that the struts can be cut to fit up against the post.

All that has to be done to produce the bevels for the strut on the left is to develop its top surface. Draw a horizontal line out from point b in the elevation, and with centre b and radius b–a describe an arc to give a′ on the horizontal line. Drop a line from a′ vertically downwards to give point a′ on the horizontal line brought out from a in the plan. Join c to a′ and b to a′ to give the bevels to apply to the top and bottom surfaces of the strut. The bevel to apply to the front and rear surfaces is seen in the elevation.

To develop the bevels for the strut on the right of the drawing, the two sides of the strut nearest the front must be developed. Draw the plan and elevation as shown, and place on the elevation a section of the strut. Draw a line through x on the section at right angles to the corners. With compass point in x and radii x–x′ and x–x′′ describe arcs to give points on the line which passes through x. From these points draw the edges of the developed sides. From points e, f, h and i draw lines outwards to intersect with the developed edges in e′, f′, h′ and i′. Join d to e′, d to i′, g to f′ and g to h′ to represent the developed sides and the required bevels.

Fig. 15 shows the plan and elevation of another post supported by a strut, the strut situated towards the front of the post. Draw the plan and elevation and put in a line above and at right angles to the strut, and step off four distances equal to the width of the sides of the strut. From a, b, c and d draw lines at right angles to the inclination of the strut to give points a′, b′, c′ and d′. The figure thus produced will give the development of the four surfaces and also the required bevels.

Hexagonal prism. Fig. 16 shows a small hexagonal prism passing through a larger square prism. The problems are to develop the portion of the smaller prism which projects from one corner of the other, and to develop the hole in the large prism through which the smaller one passes. Draw the plan and elevation. Project lines from a′, b′, c′, x, y and the end of the small prism downwards and at right angles to its elevation. On the line from the top end of the small prism, and starting from any convenient point, step off six distances equal to the widths of the sides of the small prism. From these points draw lines parallel to the small prism to intersect with the lines from a′, b′, etc., to give points a′, b′ and c′, etc. on the development. The positions of x and y are halfway across the sides in which they are situated. To develop the hole near the top of the square prism, project over to the right, and at right angles to the large prism, x, a′, b′, c′ and y, and starting from a centre line x–y, step off to the left and to the right the distances 1 and 2 seen in the plan. Vertical lines through these points will give the required points on the hole development.

Pyramids. We now come to the question of pyramids. Fig. 17 is the plan and elevation of a square box with inclined sides. Two corners

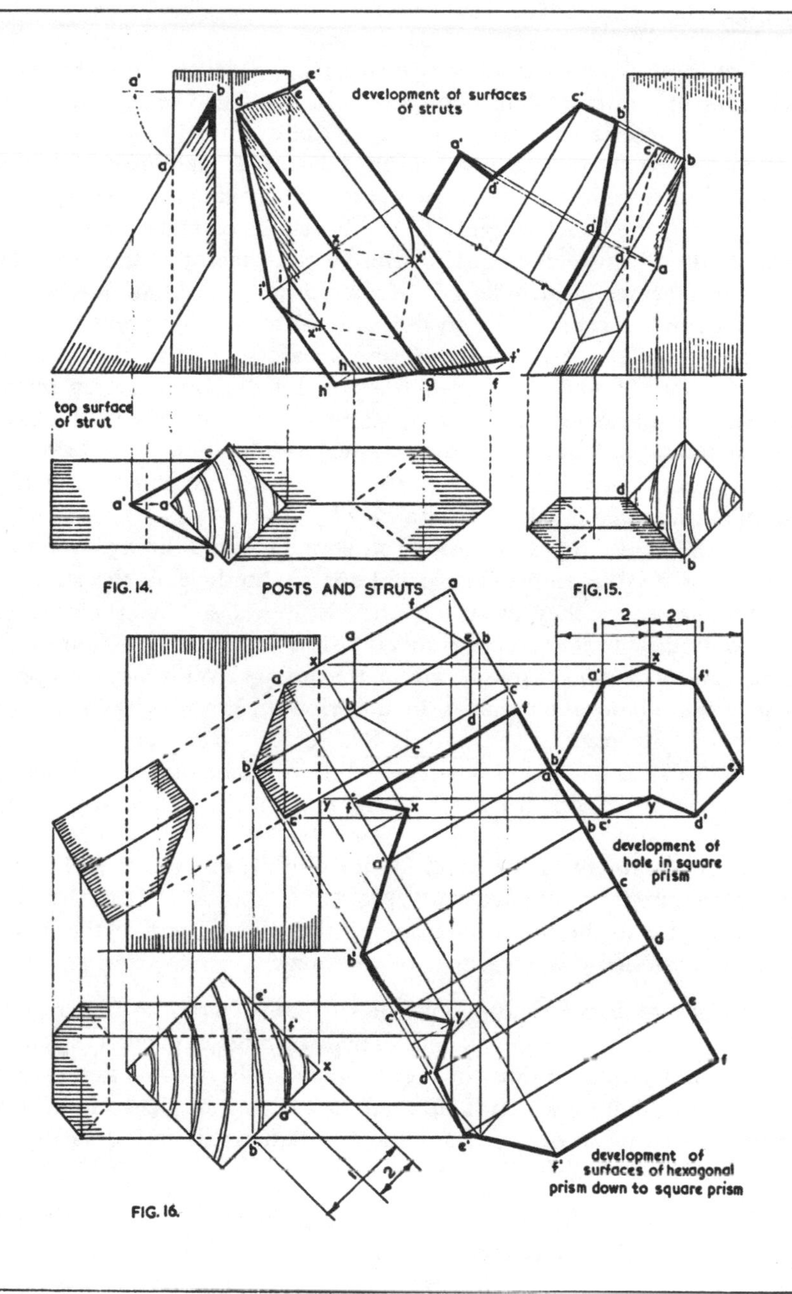
development of surfaces
of struts
top surface
of strut
FIG. 14.
POSTS AND STRUTS
FIG. 15.
development of
hole in square
prism
development of
surfaces of hexagonal
prism down to square prism
FIG. 16.

are butt-jointed, and the other two mitred. To develop the shapes of the sides of the box, with compass point in f in the elevation and radius f–d, describe an arc to give d on the extended base line. Drop a vertical line from d to give points h′ and d′ on horizontal lines brought out from h and d in the plan, f–d′–h′–g being the development.

To obtain the bevel to apply to the butt-jointed ends, draw a right angle at d and make d–e equal to the vertical height of the box. Join e to f, e–f being the true length of the corner d–f. Draw line 5–6 at any point but at right angles to d–f, and with compass point in 7 and opened to just touch line f–e describe an arc to give point 8 on d–f. Join 5 to 8 and 6 to 8. Angle 5–8–6 is the dihedral angle set up by the two sides, and angle 5–8–9 the bevel to which the ends of the sides are cut to so that the butt joints are perfect fits. The mitre bevel for the two other corners is found in the same way, but this time half of the dihedral angle is taken for the bevel.

Fig. 18 shows the top corner of some splayed linings round a window or door opening. The problems are to develop the shape of the linings where they intersect, and the bevel so that they can be butt-jointed or tongued-and-grooved. First draw the elevation of the corner and a section through one of the linings. With compass point in a in the plan, and radius a–b, describe an arc to give b′ on the horizontal line passing through a. Project upwards vertically a line from b′ to give b′ on the horizontal line brought out from b in the elevation. a–b′–c′–d is the shape of the ends of the linings where they intersect.

The development of the bevel to apply to the ends so that they can be butt-jointed or tongued-and-grooved is shown on the drawing, and is similar to the bevel for butt-jointing the corners of the box in Fig. 17. Distance y is equal to y in the plan.

Octagonal roof. Fig. 19 shows how the surfaces of a roof to an octagonal shed or kiosk can be developed. The drawings are fairly straightforward and similar to those in Fig. 17 and should need no description. The backing bevel, which is applied to the top surfaces of the hips so that a seating is provided for the tile battens, is found as for the mitre bevel for the box corners.

Oblique pyramid. Fig. 20 shows how the sides of an oblique pyramid should be developed. Remember that with a right square pyramid all four corners are equal in length, but with the oblique pyramid two

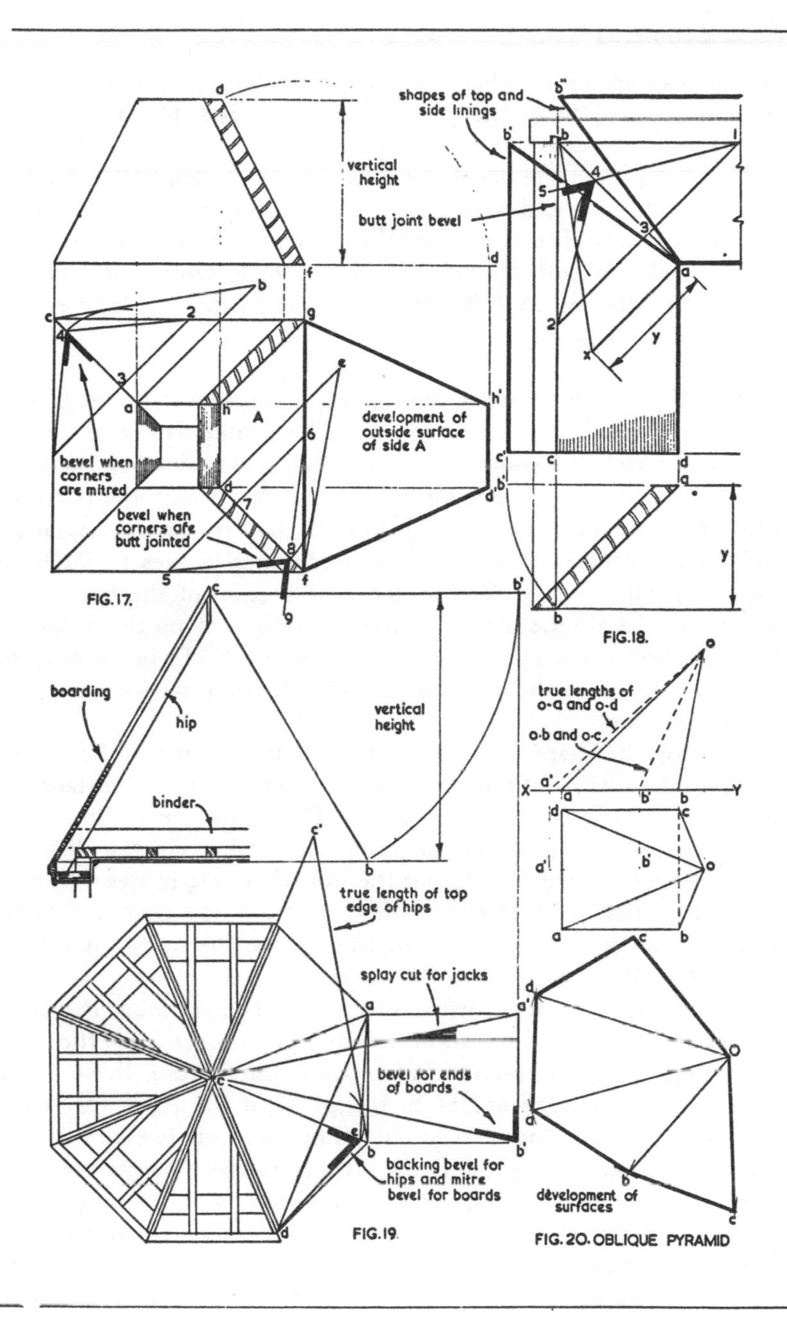

FIG. 20. OBLIQUE PYRAMID

of the corners are longer than the other two. The plan and elevation show how the lengths of the corners are developed. Having obtained these lengths the sides are developed as follows. Draw the corner o–c, making this equal to the development in the elevation. With compass open c–d in the plan and with compass point in c in the development, make an arc in the direction of d. With compass open the developed length of o–d and with its point in o in the development, make an arc to cut the first arc in d. The other points in the development can be obtained similarly.

Cylinders. Fig. 21 shows a section through a pipe or cylinder passing through a partition at an angle of 45°. The problems are to develop the shape of a clearance hole in the partition, and to develop the shape of the surface of the pipe section which passes through the partition.

The pipe and partition should first be drawn as shown, and a section of the pipe placed over that part which passes through the partition. Another circle, to represent the section of the hole in the partition, should also be placed on the drawing, making the difference in the diameter of the circles equal the amount of clearance required between the pipe and the hole surface. Divide these sections into, say, twelve equal parts.

To develop the shape of the hole, project the points on the outside circle up to the surface of the partition, and from here over horizontally to where the shape is to be developed. Draw the centre line of the development, 1′–7′, in a vertical direction, and make the distances 6′–8′, 5′–9′, etc. equal to those of the outside circle representing the hole in the partition. A freehand curve through the points obtained will give the shape of the hole to be marked on the partition surface. The shape is elliptical.

The shape of the pipe surface can be developed thus. Project to each end of the pipe the various points 1, 2, 3, etc. on the circle representing the pipe section, and draw lines from these points downwards and at right angles to the pitch of the pipe. On one of these lines and from any convenient point mark off twelve distances equal to those round the pipe. Draw lines from these points across the two sets of lines brought over from the two ends of the pipe to give two sets of intersections through which can be drawn freehand curves to obtain the shape of the pipe surface.

Fig. 22 shows a similar problem; how to obtain the shape of the

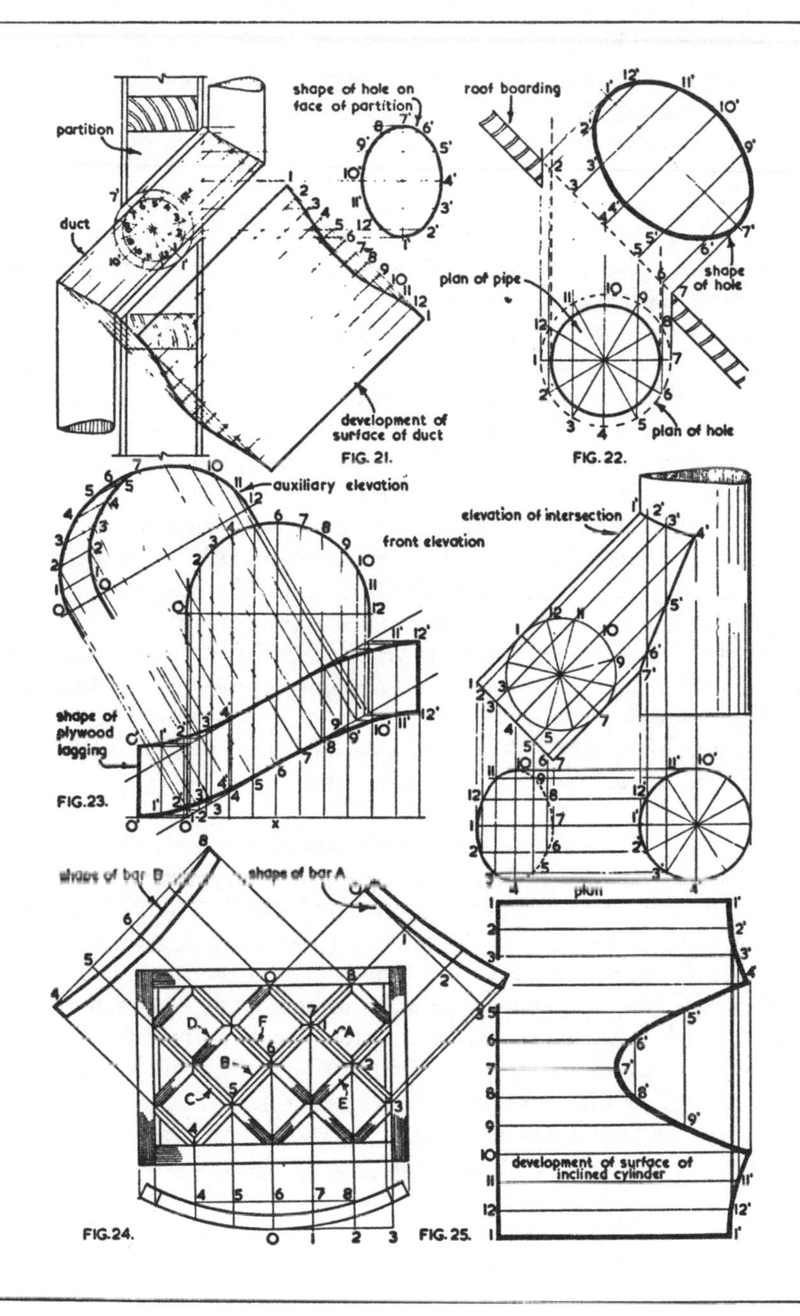

partition
duct
shape of hole on face of partition
development of surface of duct
FIG. 21.
roof boarding
plan of pipe
shape of hole
plan of hole
FIG. 22.
auxiliary elevation
front elevation
shape of plywood lagging
FIG.23.
elevation of intersection
plan
shape of bar B
shape of bar A
FIG.24.
development of surface of inclined cylinder
FIG. 25.

hole in the boarding on a roof to allow a round flue pipe to pass through. Again, the pipe should be given a clearance hole and a study of the drawings should explain how the shape of the hole is obtained.

Fig. 23 shows how the plywood lagging to a centre for a skew arch is developed. The arch, semicircular when viewed parallel with its jambs, should be divided up into, say, twelve parts, and these points projected down to each edge of the plan. On the horizontal line below the plan should be marked twelve spaces equal to those round the elevation curve, x being the centre point. Project all these points upwards vertically to intersect horizontal lines brought out from the points on each side of the plan, thus giving points on the development of the soffit of the arch, which is also the shape of the lagging required.

Fig. 24 is the plan and elevation of a sash, circular in plan with diagonal glazing bars. To obtain the shape of glazing bar A, project points 0, 1, 2 and 3 downwards to the horizontal line just below the plan, giving points 0, 1, 2 and 3 on this line. Also project lines from 0, 1, 2 and 3 over and at right angles to the bar in the elevation, and draw the line 0–3 parallel to the bar in the elevation. Make the distances between the line 0–3 and the development of the bar equal to those seen in the plan. This shape can then be used as a templet for the bars marked A, C, D and E. The shape of the templet for bars B and F is found similarly as shown in the drawings.

Intersecting cylinders. Another exercise involving the cylinder which often proves valuable is shown in Fig. 25. The top drawings show the plan and elevation of two intersecting cylinders. The problem here is to develop the shape of the surface of the inclined cylinder so that it will fit exactly over the vertical one.

Draw the plan and elevation and divide the surface of each cylinder into, say, twelve equal parts. Project the various points on the plan of the inclined cylinder over to the plan of the vertical one to give points 1′, 2′, 3′, etc. From here they should be projected upwards to intersect with the lines taken along the elevation of the inclined cylinder to give 1′, 2′, 3′, etc. A freehand curve through the points thus obtained will give the elevation of the intersection.

To develop the surface of the inclined cylinder, draw the vertical line 1–1 seen below the plan, and mark off twelve distances equal to those round the surface of the inclined cylinder. Draw horizontal lines from these points and make these lines equal to those on the surface. These lengths can be obtained from the elevation. Freehand curves through the ends of the lines will give the surface development.

Cone problems. Fig. 26 is the plan and elevation of a cone. In the companion volume *(Practical Carpentry and Joinery)*, it was shown how to develop the surface of the solid, and also how to develop an elliptical section through the cone. Fig. 26 explains the method of developing two more sections, namely, the parabola and hyperbola. If the section line is parallel to one of the edges of the elevation, x–7′, a parabolic section will be produced, when the line is vertical, but not on the centre line of the elevation, a–7″, a hyperbolic curve will be the shape of the section.

To produce the parabolic section, draw the section line parallel to one edge of the elevation, and drop lines down vertically from points 3′, 4′, etc. to give a plan of the section. Then draw lines at right angles to the section line in the elevation from all the points on line x–7′, and, starting from a centre line z–7′, make the distances across the development equal those across the plan of the section.

To produce the hyperbolic section, draw line a–7″, being careful not to position it on the centre line of the solid. Project the line down into the plan to obtain the plan of the section. This is also a straight line. From the various points on the plan project lines horizontally over to where the section is to be produced, and on the centre line c–7″ mark off the various heights taken from the elevation; for example, make c–d in the development equal a–5″ in the elevation, c–e equal a–6″, and c–7″ equal a–7″. Project lines parallel to the base of the development to give 5″, 6″, 8″ and 9″. A freehand curve through these points will give the development of the hyperbolic section.

Parabola. Figs. 27 and 29 show two more methods which can be used for setting out the parabola. When the height and the width are known the method shown in Fig. 27 can be used; when the positions of the directrix and the focal point are known the method shown in Fig. 29 is used. When using this second method it must be remembered that the distances between the directrix and the curve, and the focal point and the curve is always in the ratio of 1:1.

If it is necessary to find the position of the focal point of a parabola the method shown in Fig. 28 should be followed. Construct the parabola as in Fig. 27 and construct a tangent at any point p. Mark point p anywhere on the curve and draw a horizontal line from it over to the centre line. With compass point in b and radius a–b draw the semicircle to give a′ on the centre line. Draw a line from a′ to p. This is a tangent to the curve at p. To produce the focal point draw a vertical line upwards from p, and, with p as centre and radius any convenient distance, draw the arc d–g, cutting the tangent in e. With compass point in e and radius e–d draw an arc to give point g on d–g. Draw a line from g to p, cutting the centre line in F. This is the focal point of the parabola.

If a number of normals are required from various points on the parabolic curve, say, from p′ and p″, place the compass point in F and with radii F–p′ and F–p″ draw arcs to give x′ and x″ on the centre line. Lines through x′–p′ and x″–p″ will give the required normals.

Pitched and conical roof intersections. Fig. 30 is the plan and elevation of a conical roof intersecting a pitched roof. The main roof is pitched at the same angle as the rafters of the conical roof. The problem is to develop the shape of the curb required to be placed on the main roof so that the ends of the rafters on one half of the conical roof will have a seating, and to which they can be fixed. This problem is similar to producing a section through a cone.

The section being parallel with the edge of the cone, the shape of the section (or curb, as it is in this case) is parabolic. Compare this drawing with Fig. 26.

Fig. 31 is the plan of the roof of a circular bay window abutting the wall of a house. The width of the roof from left to right is x–y, and the distance it stands out from the wall is z–5. The problems here are to produce the shape of the wall piece which is hyperbolic, and to which the tops of the rafters will be fixed; to develop the shapes of some of the rafters; and develop the surface of one half of the roof.

Having drawn the plan of the bay roof it is necessary to draw the plan and elevation of the cone of which the bay roof is a part. The edges of the elevation of the cone must be drawn at an angle equal to the pitch of the bay roof. In this case the pitch is 45°. Next draw the plan of the wall piece and divide the curve of the wall plate into a number of parts, say eight. Number these 1, 2, 3, etc. Draw lines from these points, which are on the base of the cone, up to the top

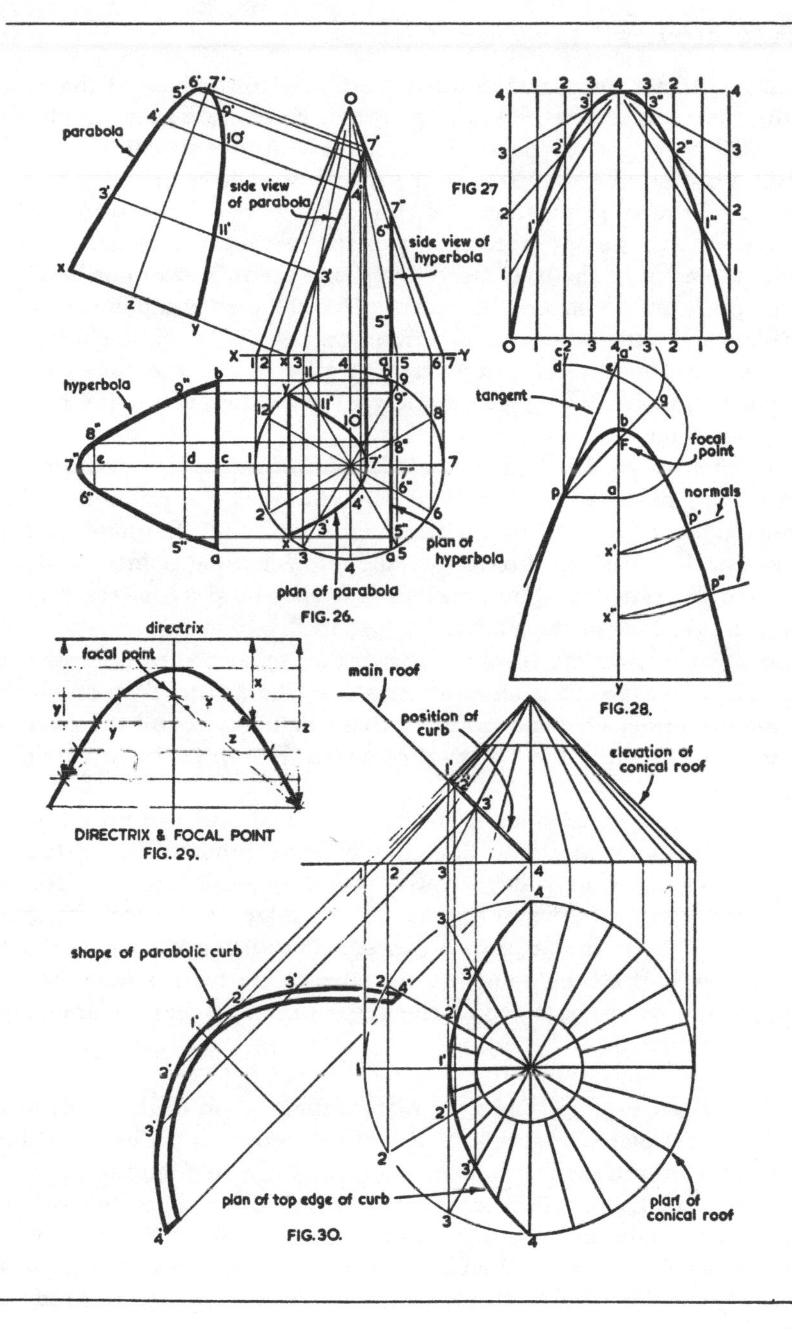
parabola
side view
of parabola
side view of
hyperbola
hyperbola
plan of
hyperbola
plan of parabola
FIG. 26.
FIG 27
tangent
focal
point
normals
FIG.28.
directrix
focal point
DIRECTRIX & FOCAL POINT
FIG. 29.
main roof
position of
curb
elevation of
conical roof
shape of parabolic curb
plan of top edge of curb
plan of
conical roof
FIG.30.

point w. Project the points upwards vertically to the base of the cone in the elevation to give points 1, 2, 3, etc. From here they should be projected up to the top point of the cone, w, in the elevation.

Lines projected upwards from the plan from where the lines on the surface of the cone pass through the front edge of the wall piece, b, c, d, etc., to the appropriate line in the elevation, give the outline of the top edge of the wall piece. Only one point, e, will not be produced thus, and to obtain this point, place the compass point in w in the plan, and with radius w–e describe an arc to give e′ on the horizontal line from w. Project e′ upwards to give e′′ on the edge of the elevation, and from here horizontally over to the centre line of the elevation to give point e.

The method of developing two of the roof rafters is fairly simple and should be well understood by the reader. Take rafter B. With compass point in w swing all the points on rafter B round to the horizontal line passing through w, and project these points upwards to where the rafter is to be developed. Make the pitch of the rafters equal to the slope of the elevation edge, in this case 45°.

To develop half the surface of the roof, open the compasses the length of one edge of the elevation and describe an arc. Step off along the arc a number of spaces equal to those half way round the plan of the roof and number these 1, 2, 3, etc. Join these points with straight lines to w.

The true distances from w down to b, c, d and e must now be marked on the appropriate line in the development. For instance, the true distance from w down to b is found by projecting a horizontal line from b in the elevation across to the edge of the cone to give point b′. w–b′ is the distance required. The distance from w down to e is w–e′ . If all the points on the surface are treated thus, points b, c, d and e will be obtained on the development so that the drawings can be completed.

Conical turret. Fig. 32 shows the plan and elevation of the walls of a conical turret placed centrally on a roof pitched at 45°. The problems are to develop the shape of one curb on the main roof, and to develop the shape of the plywood required to cover the wall of the turret. These developments should be straightforward for the reader who has studied the foregoing drawings on the cone and should need little explanation. The curb is elliptical in shape, of course, and to produce

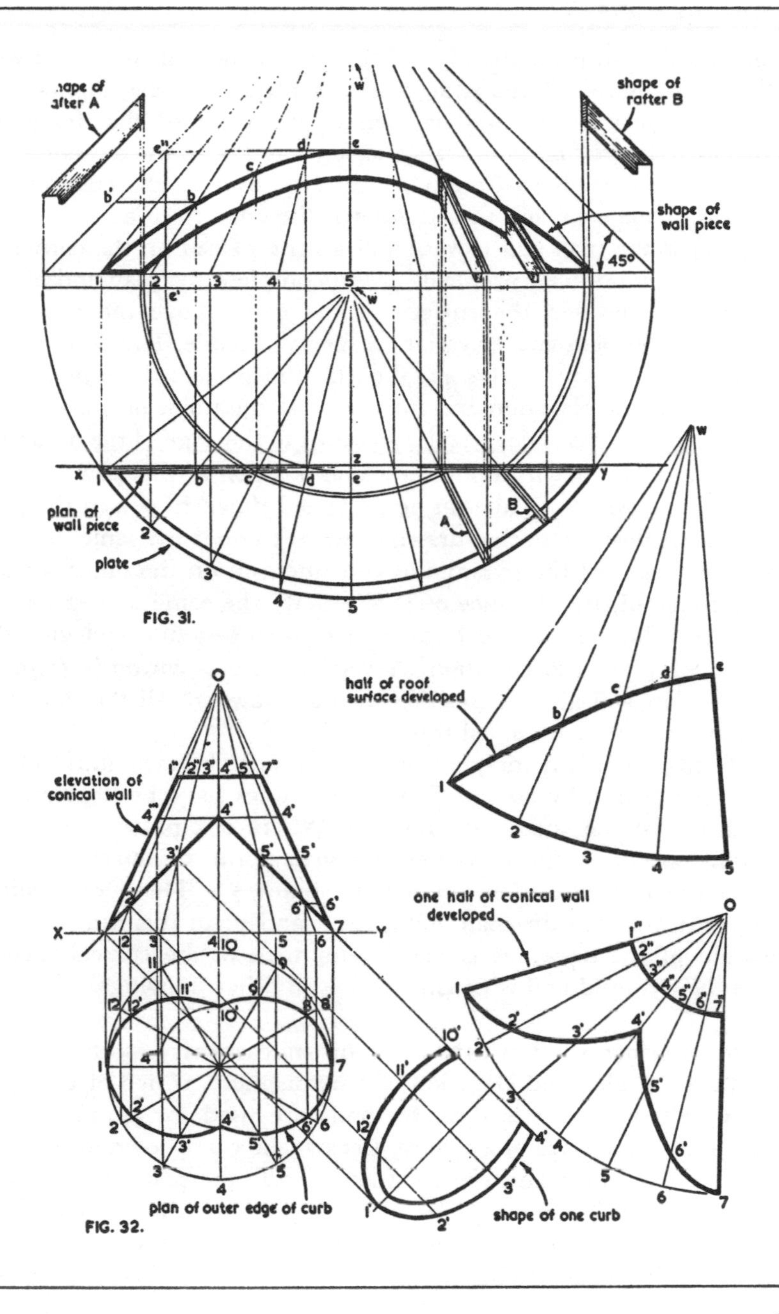
shape of rafter A
shape of rafter B
shape of wall piece
45°
plan of wall piece
plate
half of roof surface developed
elevation of conical wall
one half of conical wall developed
plan of outer edge of curb
shape of one curb
FIG. 31.

FIG. 32.

it one has first to draw the plan of the intersection of the turret wall with the main roof. This is seen in the plan. The development of half the surface of the wall is seen to the right of the drawings. Remember that to find points 2′, 3′, 4′, 5′, and 6′ in the development one has to project lines from 4′, 5′ and 6′ over to the edge of the elevation of the cone and measure down the edge from o.

Fig. 33 is the plan and elevation of a cone placed on the corner of a square pyramid. The problems are to complete the plan and elevation, and to develop the surface of the cone above the pyramid. First draw the plan and elevation as far as possible. Divide the plan of the cone into, say, twelve equal parts, and project these points up to its base in the elevation and from here to the point of the solid.

It will be seen that lines 1 and 4 go down to the edge of the pyramid, and line 2 comes down to point 2′. The position of this point can be seen in the elevation. This can be projected down to o–2 in the plan to obtain its position in that drawing. As o–3 line is the same distance from the corner of the pyramid as o–2 line, but on the other surface of the pyramid, the distance o–3′ is exactly the same as o–2′. Consequently a horizontal line from 2′ across to o–3 line will give the position of 3′ in the elevation. A vertical line downwards from 3′ into the plan will give its position in that drawing. All the points in the drawing can be obtained thus.

Only two points, namely x′ and y′, are not obtained thus. These can be positioned by assuming two more lines have been placed on the surface of the cone, o–x and o–y. Where o–x passes across the pyramid corner in the elevation will give point x′ and where o–y crosses the corner farther up will be point y′. These two points dropped down into the plan will give x′ and y′ in this drawing. To the right of the drawings is the development of the conical surface above the pyramid and is constructed as in other drawings.

Sphere. Another solid which crops up in carpentry is the sphere. Fig. 34 is the plan and elevation of a hemisphere, which of course is exactly half of a sphere. Two methods are used for approximately developing the surface of a sphere. Let us take that shown in the plan first. Draw the plan and elevation and divide the plan up into, say, six equal parts. To develop one of these sections, divide half of the elevation curve also into six equal parts and drop these points down and across the section to be developed. Project the edges of the section

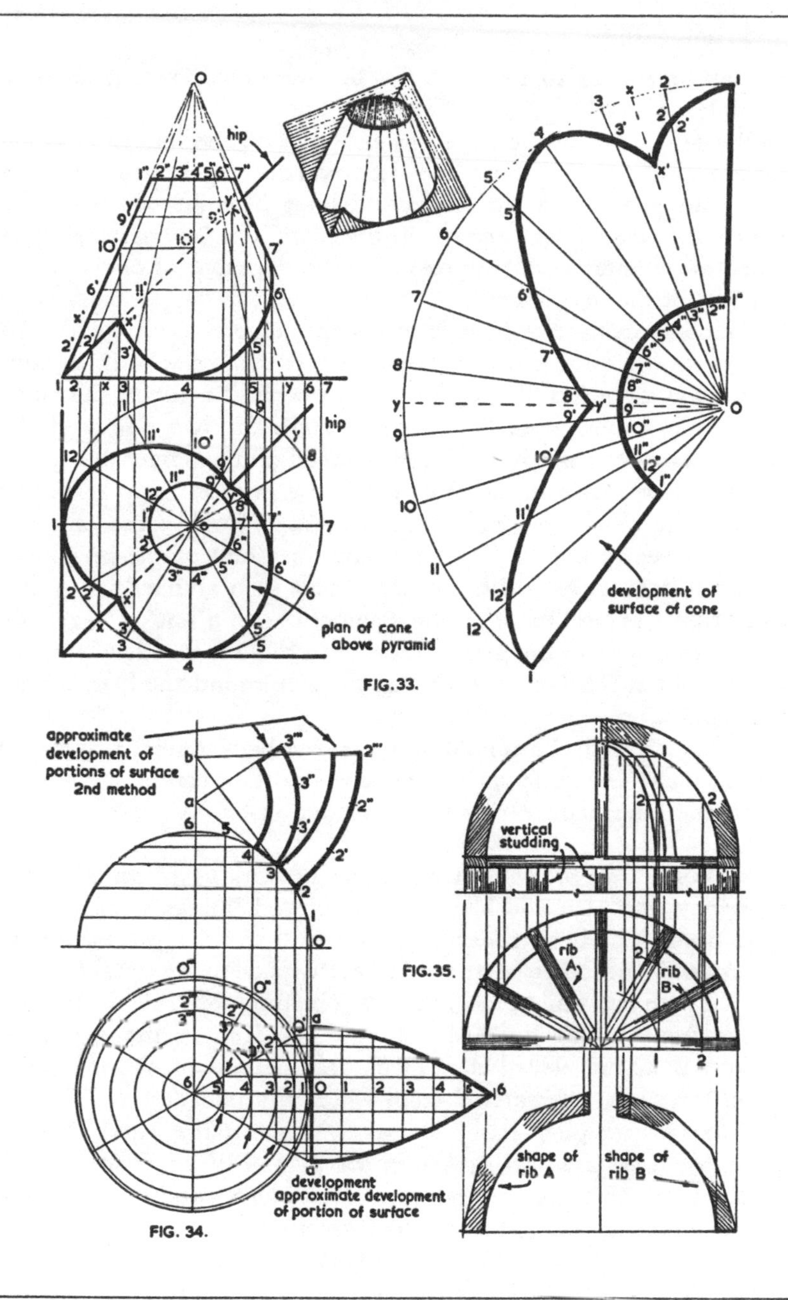

FIG. 33.

FIG. 34.

FIG. 35.

to be developed out to intersect the line brought down from o to give points a and a′.

On the centre line of the development mark off the six spaces equal to those round the elevation, and draw vertical lines through these points to intersect with the horizontal lines brought out from the points on the edges of the section in the plan. The intersections thus obtained are points on the approximate development of one-sixth of the surface of the hemisphere.

The second method, which is also approximate, is to divide the surface of the hemisphere into horizontal strips. Each strip is considered to be a portion of the surface of a cone. To obtain the point of each cone, a line must be drawn through the two points at the extreme edge of the portion being developed and extended up to the centre line. For instance, let us consider the portion between points 2 and 3 in the elevation. Draw a line from 2 through 3 and onwards to meet the centre line in b. This is the top point of the imaginary cone and b–2 its edge. With compass point in b and radii b–3 and b–2 describe arcs, and mark off the distances 2–2′, 2′–2′′, 2′′–2′′′ on the arc from 2 seen in the plan. Join 2′′′ to b. This is a development of the strip of surface one-fourth the distance round the hemisphere at that height.

Fig. 35 is a practical example based on the hemisphere, and shows a wooden niche frame. It shows how the elevation can be completed and how the intermediate ribs are developed.

Conical linings. Fig. 39 is another example of work based on the cone. It represents the plan and elevation of splayed linings with a semicircula. head. The head consists of two semi-circular pieces, one at the front and the other at the rear, each made in two halves and connected with handrail bolts. They are joined by four rails which run from the front to the back of the head, as seen in the elevation. At A and B in the elevation is given the method of obtaining the width of the pieces of material from which the rails are cut. Fig. 40 shows the two templets required for cutting out the rails. The dimension for these are obtained from the plan. Fig. 41 illustrates the application of the templets and the bevels, also obtained from the plan, enabling the rails to be shaped correctly.

To obtain the shapes of the thin plywood panels, the face of the panel in the plan should be extended to the centre line to give point

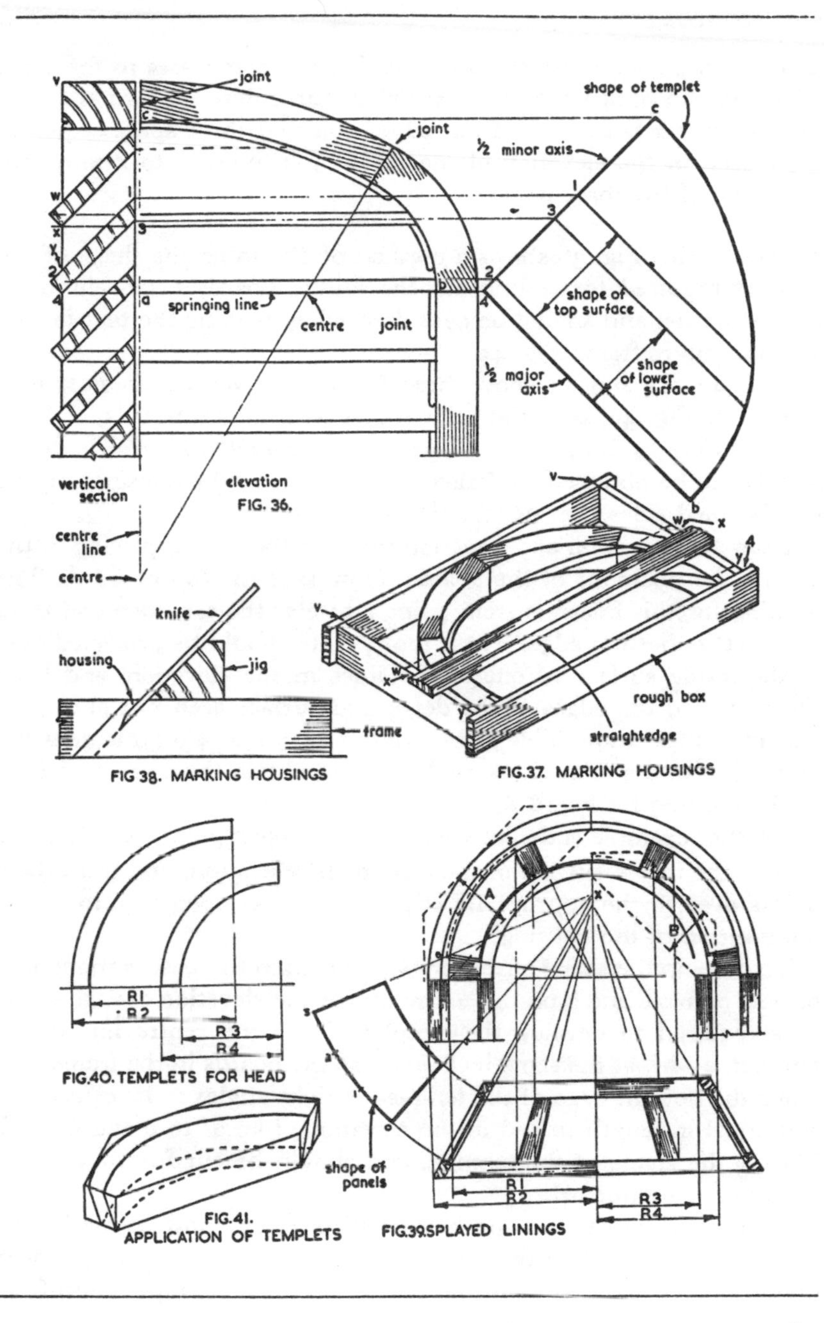

FIG. 36.

FIG 38. MARKING HOUSINGS

FIG.37. MARKING HOUSINGS

FIG.40. TEMPLETS FOR HEAD

FIG.41. APPLICATION OF TEMPLETS

FIG.39. SPLAYED LININGS

x, and using x for a centre, and opening the compasses to the front edge and the back edge of the panel in turn, describe two arcs. On that which starts from the front edge, mark off three spaces equal to those seen in the elevation of the panel. Join point 3 to x to obtain the shape of the three panels.

Louvre boards. Fig. 36 shows a method of obtaining the shape of the templet required for setting out the louvre boards to an elliptical-headed frame, and should be consulted when reading the text for the circular louvre frame Fig. 44.

Figs. 42, 43 and 44 show three frames for ventilation purposes. The first, Fig. 42, is a triangular frame with louvre boards, and the problems are to develop the shape of the templet from which all the boards can be obtained, and also the positions of the housings on the two inclined sides. Let us take the latter first.

Draw the elevation and a section through the frame, placing in the section the positions of the boards. Now take the top board A. The housing for this board is from point 1 to e at the top edge and from 3 to 4 at its bottom edge. These four points should be projected over to the inside surface of one of the sides in the elevation, and from there over to the edges of the developed surface seen just above the side of the elevation. This gives points 1, e, 3 and 4 on this drawing. Join 1 to e and 3 to 4 with straight lines to obtain the correct position of the housing for board A.

Another example is the louvre board C. Project points w, x, y and z over to the inside surface of the side in the elevation, and from these points over to the appropriate edge of the developed side to obtain the position of the housing.

To develop the shape of the templet, which in this case is triangular, project points a and b on the centre line of the elevation over to where it is required to develop the templet. Draw the centre line of the templet, a′–b′, at the same inclination as the boards in the frame, and make the bottom edge of the templet at right angles to its centre line and equal in length to c–d in the elevation. Join a′ to c and a′ to d. This is the shape of the templet, and should be cut from a piece of plywood or hardboard.

The shapes of the top and bottom surfaces can now be marked on the templet. To obtain the shapes of the surfaces of board A, project points 1, 2, 3 and 4 over to the centre line of the templet, and from

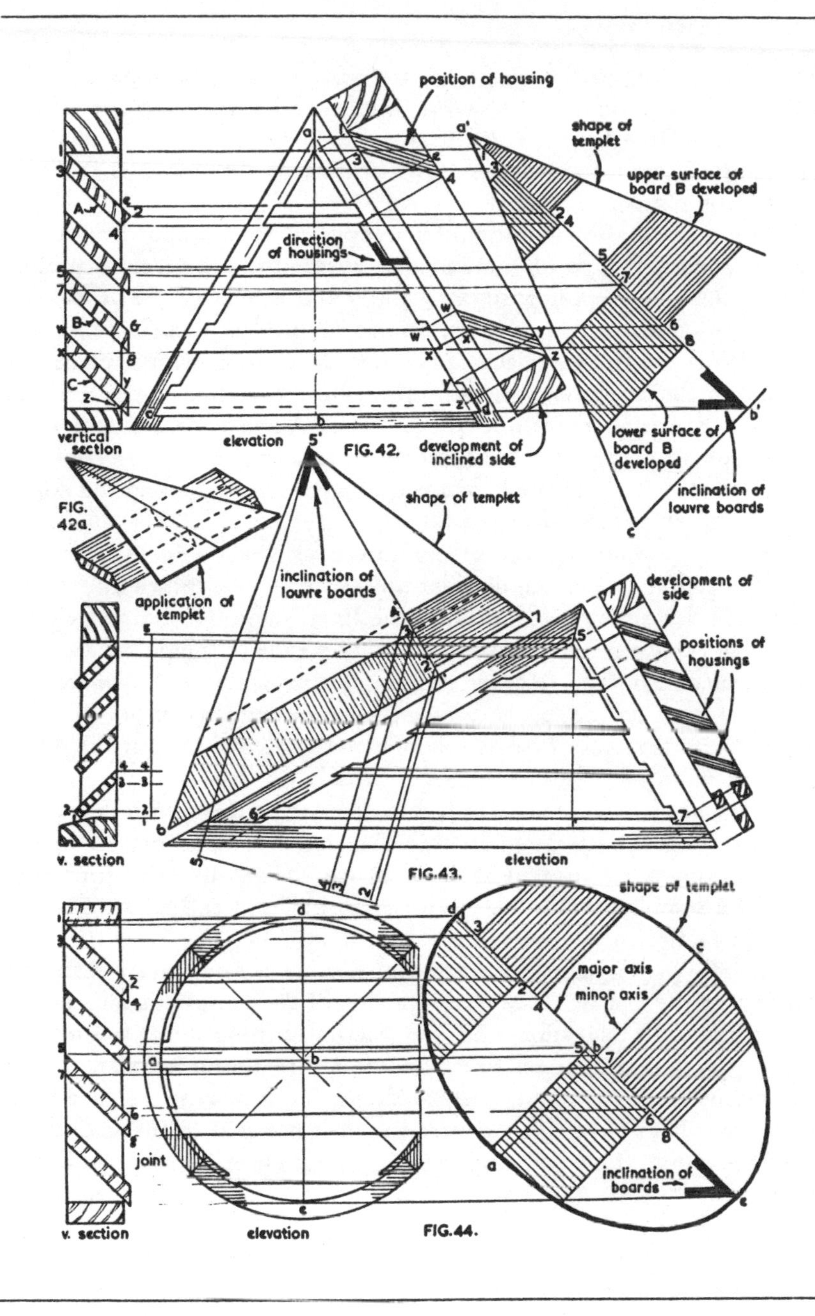
position of housing
shape of templet
upper surface of board B developed
direction of housings
vertical section
elevation
FIG. 42.
development of inclined side
lower surface of board B developed
inclination of louvre boards
FIG. 42a.
application of templet
shape of templet
inclination of louvre boards
development of side
positions of housings
v. section
elevation
FIG. 43.
shape of templet
major axis
minor axis
joint
inclination of boards
v. section
elevation
FIG. 44.

these points over to the edges of the templet at right angles to the centre line. The distance between lines 1 and 2 is the shape of the top surface of the board, and the distance between points 3 and 4 is the shape of the lower surface. The shapes of the board B are also seen on the templet.

Fig. 42a shows how the templet is applied to the board material. Suppose board B is to be marked out. First obtain a piece of timber long enough for the purpose, and apply the bevels to the front and rear edges. Square a centre line round all the four surfaces of the board. With the front surface of the board facing upwards, place the templet on the board, making sure that the centre line of the templet is directly over the centre line of the board. Also ensure that the lines 5 and 6 on the templet are directly over the top and lower edges of the board. The ends of the board can now be marked along the edges of the templet directly over the board.

Now turn the board over so that the back surface faces upwards. Place the centre line of the templet directly over the centre line of the board, but this time make sure that the lines 7 and 8 are directly over the top and bottom edges of the board. The ends of the board are now marked again and the templet removed. The ends of the board are cut to the lines on the top and lower surfaces and then fitted into the frame. Treat the other boards in the same way, making sure that the correct lines on the templet are used for each board.

The second triangular louvre frame, Fig. 43, is dealt with in exactly the same way. The templet in this drawing has been placed in the position shown because of lack of space. Normally, the line 1–5, below the templet, would be in the same position as line 1–5 just to the right of the vertical section.

Although the circular louvre frame, Fig. 44 is an entirely different shape from the last two, the development of the templet is almost the same. Draw the elevation and the section through the frame and project points d and e over to where the templet is to be drawn. The centre line of the templet, d–e, is drawn at the same angle as the louvre boards and the other centre line a–c is equal to the width a–c in the elevation. The shape of the templet is elliptical and the board shapes are placed on the templet in the same way as for the triangular louvre frames.

The drawings in Fig. 36 should be fairly straightforward because the geometry for this semi-elliptical headed louvre frame is the same as for the circular frame. Only half of the required templet is shown. To mark the housings on the curved inside surface of these frames take the frame in Fig. 36 as an example. A rectangular box should be made, Fig. 37, so that the curved head just fits into it. On the sides of the box from point v should be marked the distances seen in the section, Fig. 36, i.e. v–w, v–x, v–y, etc. A straight edge placed across each pair of points, w–w, x–x, etc. enables the positions and direction of the housings to be marked on the inside surface of the frame, see Fig. 37. The housings are marked by placing a jig, Fig. 38, across each pair of lines, and marking the slope of the housings with a piece of steel sharpened to a knife edge. If the jig bevel is the same as the inclination of the louvre boards, and the knife is kept flat on the jig, as in Fig. 38, the housings marked on the inside surface of the frame will be at the correct angle.

Raking mouldings. When two horizontal mouldings intersect at a corner in a building the sections of the mouldings are exactly the same, but if one or both of the mouldings were inclined, their sections would have to be different.

Take the simple case shown in Fig. 45. The plan shows three mouldings, A, B and C. Let us assume that moulding A is horizontal, B is inclined at 30°, and C is horizontal. To develop the shapes of B and C when A is known, draw a section of A immediately above its plan as shown, and from all the points on the moulding draw lines upwards at the inclination of moulding B. Draw o–o′ the back edge of moulding B, at right angles to the inclined lines representing the elevation of the moulding. Set off along the top line o–1–2–3 equal to the thickness of mould A, and drop lines across the elevation parallel to the back edge o–o′. The various intersections will give the shape of moulding B. If the inclined lines from moulding A are carried upwards to where moulding C is to be developed, the intersections will again be obtained by making the distances across the moulding equal to those on mouldings A and B, and dropping lines vertically to meet the inclined lines.

Mitre bevels. To develop the mitre bevels, take moulding A first. To obtain the bevel so that it mitres with B correctly, its top surface must be developed. With compass point in o and radius o–x describe an arc to give x′ on the horizontal line o–x′. Drop a line down to the plan to meet a horizontal line brought out from x to give x′

in this drawing. The bevel required to apply to the end of A is indicated at x′.

To obtain the bevels to apply to the ends of moulding B, its top surface must be developed. This is shown in the elevation. With compass point in o on the section of mould B, and with radius o–y describe an arc to give point y′ on the extended back edge of the section. Draw a line through y′ parallel to the inclined moulding to meet lines brought up at right angles to the inclination to give points x′′ and z. The bevels are also indicated. The bevel to apply to the end of mould C is seen in the plan because the top surface of the moulding is in the horizontal plane.

Fig. 46 shows the plan and elevation of two mouldings intersecting at an angle of 135°. Mould A is horizontal and B is inclined at 30°. The shape of mould A is known. Draw the plan of the two mouldings and the elevation of the horizontal moulding A. Note that when an inclined moulding is involved with a horizontal moulding, the inclined moulding should be drawn parallel to the top and bottom edges of the drawing sheet. Draw a section of mould A on the elevation, and mark the various distances 0–1–2–3 across its thickness on the plan of the moulding. Project these points on the plan over to the mitre and upwards vertically to meet the horizontal lines brought over from the section in the elevation, so that an elevation of the mitre can be constructed. From all the points on the mitre elevation draw lines parallel to the inclination of moulding B and draw the back edge of this, o–o′, at right angles to the inclined lines.

Make the distances across moulding B equal to those across A, 0–1–2–3, and drop lines from these points parallel to the back edge to obtain a series of intersections. Join these up as shown to obtain the shape of moulding B.

Fig. 48 shows the intersection of two mouldings, A and B, both inclined at 30°, the shape of moulding A being known. Draw the plan of the two mouldings and the elevation of mould A. Note that when two inclined mouldings are involved, the plan of the moulding of known shape is drawn parallel to the top and bottom edges of the drawing sheet. Draw the section of the known moulding and mark off the distances seen in the section across its plan. Project the points across the plan over to the mitre line, and upwards vertically to meet the inclined lines brought up from the elevation section. These intersections will give an elevation of the mitre.

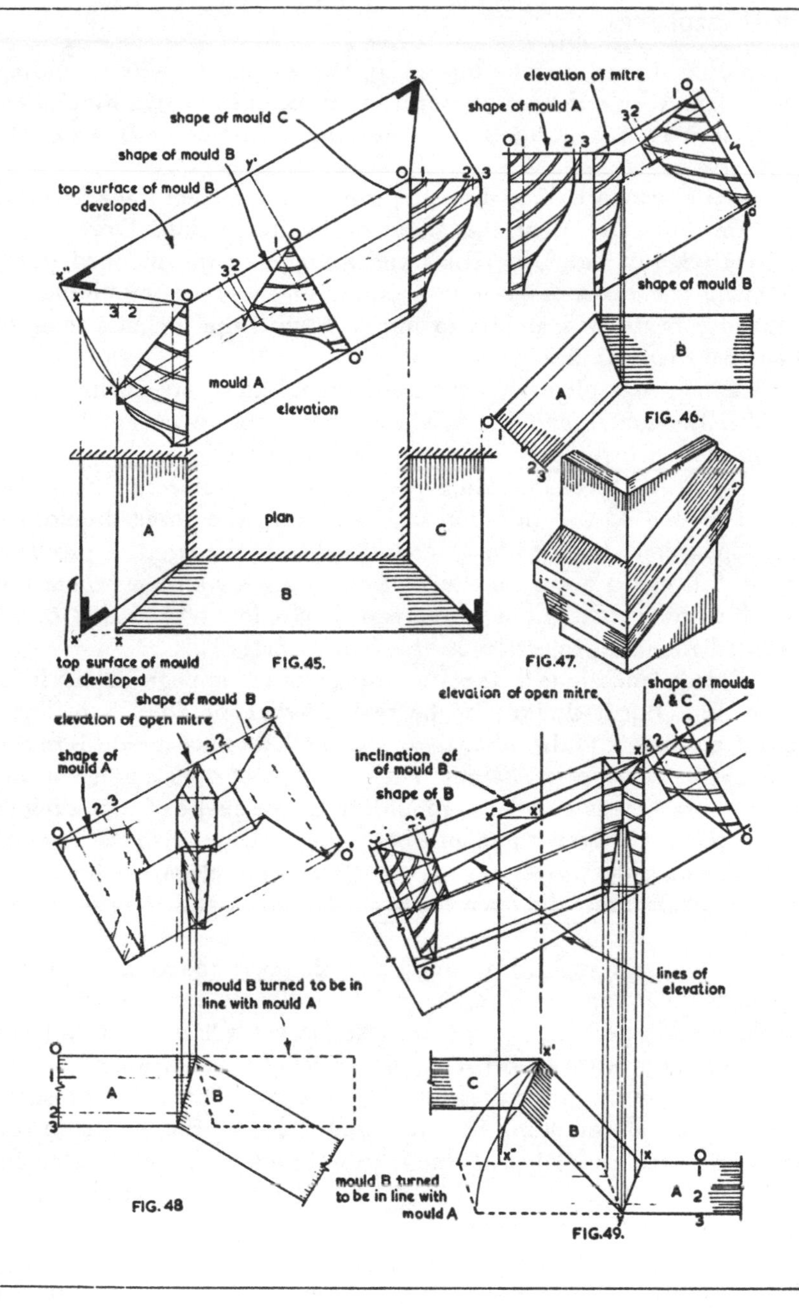
shape of mould C
shape of mould B
top surface of mould B developed
mould A
elevation
plan
A
B
C
top surface of mould A developed
FIG.45.
elevation of mitre
shape of mould A
shape of mould B
FIG. 46.
FIG.47.
shape of mould B
elevation of open mitre
shape of mould A
mould B turned to be in line with mould A
FIG. 48
elevation of open mitre
shape of moulds A & C
inclination of of mould B
shape of B
lines of elevation
mould B turned to be in line with mould A
FIG.49.

Moulding B must now be turned so that it is in line with moulding A, see plan. If this is done the mitred surface of the two mouldings will be seen in the elevation, and the mitred surface of B is exactly the reverse of A.

Draw the second half of the open mitre, and project lines from all the points of mitre B at the angle of its inclination. Draw o–o′, the back edge of moulding B, at right angles to the inclined lines, and make the distances o–1–2–3 equal to those on moulding A. Drop lines from these points to intersect with the inclined lines to obtain the shape of moulding B.

Fig. 47 shows a pictorial view of two mouldings, one inclined and the other horizontal, and clearly shows that if the two mouldings are to intersect correctly, they must have different sections.

Fig. 49 shows three mouldings, A, B and C, intersecting as in the plan. The line of the elevation indicates that the three mouldings must appear in a straight line when viewed from the front. A problem such as this could occur on the wall adjoining a staircase where the mouldings follow the pitch of the stairs. Mouldings A and B could be mitred round a projection on the wall surface.

As already mentioned, the mouldings must appear to be in a straight line, but it should also be realized that mould B is inclined at a different angle to the other two. The inclination of mould B must, therefore, be developed. Assume that the shape of mould C is known. As moulding A runs parallel to moulding C, the shape of moulding A is the same as C. Draw the plan to any given shape; also the lines of the elevation. Add the section of A between the elevation lines and above its plan. Draw the open mitre of moulds A and B as shown in Fig. 48.

To develop the inclination of mould B, place the compass point in x, in the plan, and with radius x–x′ describe an arc to give x′′ on the horizontal line brought out from x. Project a line upwards from x′′ to intersect with a horizontal line brought out from x′ in the elevation to give x′′ in the drawing. Join x′′ to x. This line represents the inclination of moulding B. Draw lines from all the points on the open mitre of B parallel to its inclination line x–x′′, and develop its shape between these lines as described in the other drawings and as indicated to the left of the elevation.

Oblique planes. These and their development play a big part in the

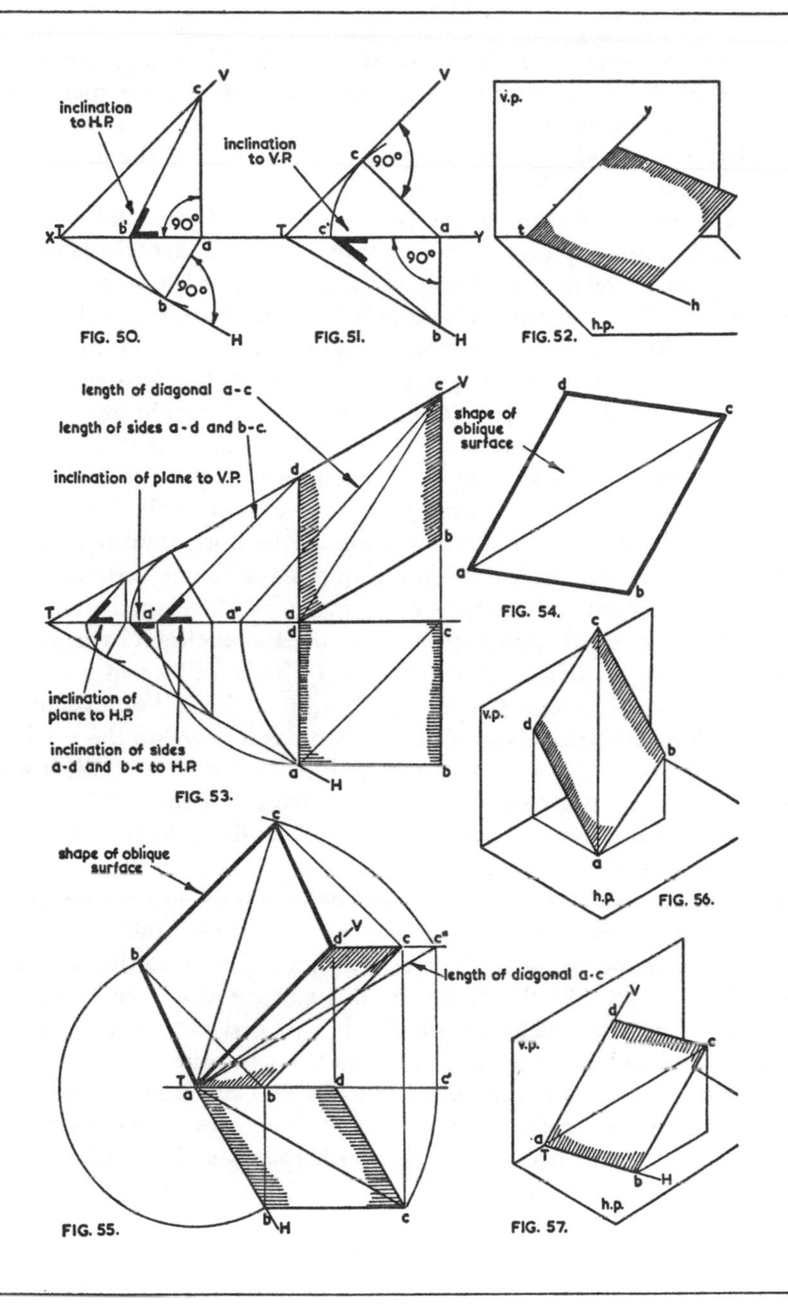

inclination to H.P.
inclination to V.P.
90°
v.p.
h.p.
FIG. 50.
FIG. 51.
FIG. 52.
length of diagonal a-c
length of sides a-d and b-c.
inclination of plane to V.P.
shape of oblique surface
inclination of plane to H.P.
inclination of sides a-d and b-c to H.P.
FIG. 53.
FIG. 54.
FIG. 55.
FIG. 56.
FIG. 57.
length of diagonal a-c

life of the carpenter and joiner. Roof work, hand-railing, and other forms of work often call on the development of oblique planes for the correct setting out of such work. Fig. 52 illustrates what an oblique plane looks like, and Figs. 50 and 51 show how to find the inclination of such a plane to the horizontal and vertical planes.

Let V.T and H.T, Fig. 50, be the vertical and horizontal traces of an oblique plane. To find its inclination to the horizontal plane it is necessary to develop the shape of a right-angled triangle placed beneath the plane so that its base a–b is at right angles to the H.T. and its hypotenuse b–c, is in contact with the surface of the plane. To develop the triangle let a–b be the base of the triangle a–b–c. a–b can be placed anywhere along the H.T. and at right angles to it. From a draw a vertical line to meet the V.T. in c. This is the second side of the triangle. With compass point in a and radius a–b describe an arc to give b′ on the x–y line. Join b′ to c. b′–c is the hypotenuse of the triangle and a–b′–c its inclination to the horizontal plane.

To obtain the inclination of an oblique plane to the vertical plane, Fig. 51, a right-angled triangle must again be developed, this time with its base at right angles to the x–y line but inclined over so that the side a–c is at right angles to the V.T. This will again make the hypotenuse b–c in contact with the oblique surface. To develop the triangle place the side a–b anywhere so long as it is 90° to the x-y line. From a, and at right angles to the V.T., draw side a–c. With a as centre and radius a–c describe an arc to give c′ on the x–y line. Join c′ to b. The developed triangle is a–b–c′, and angle b–c′–a is the inclination of the oblique plane to the V.P.

In Fig. 53, a–b–c–d is the plan and elevation of an oblique plane, and Fig. 56 a pictorial view of the surface. The problems are to develop the shape of the surface and its inclinations to the V.P. and the H.P. To develop the shape of the surface draw the plan and elevation. The true length of sides a–b and c–d can be seen in the elevation because they are parallel to the V.P., but the lengths of sides a–d and b–c must be developed. If we develop the length of a–d this will also give the length of b–c because they are equal. Place the compass point in d in the plan, and with radius a–d describe an arc to give a′ on the x–y line. Join a′ to d in the elevation to obtain the lengths of a–d and b–c.

We now require to develop the distance between two opposite corners, say a and c. To find the length of diagonal a–c place the

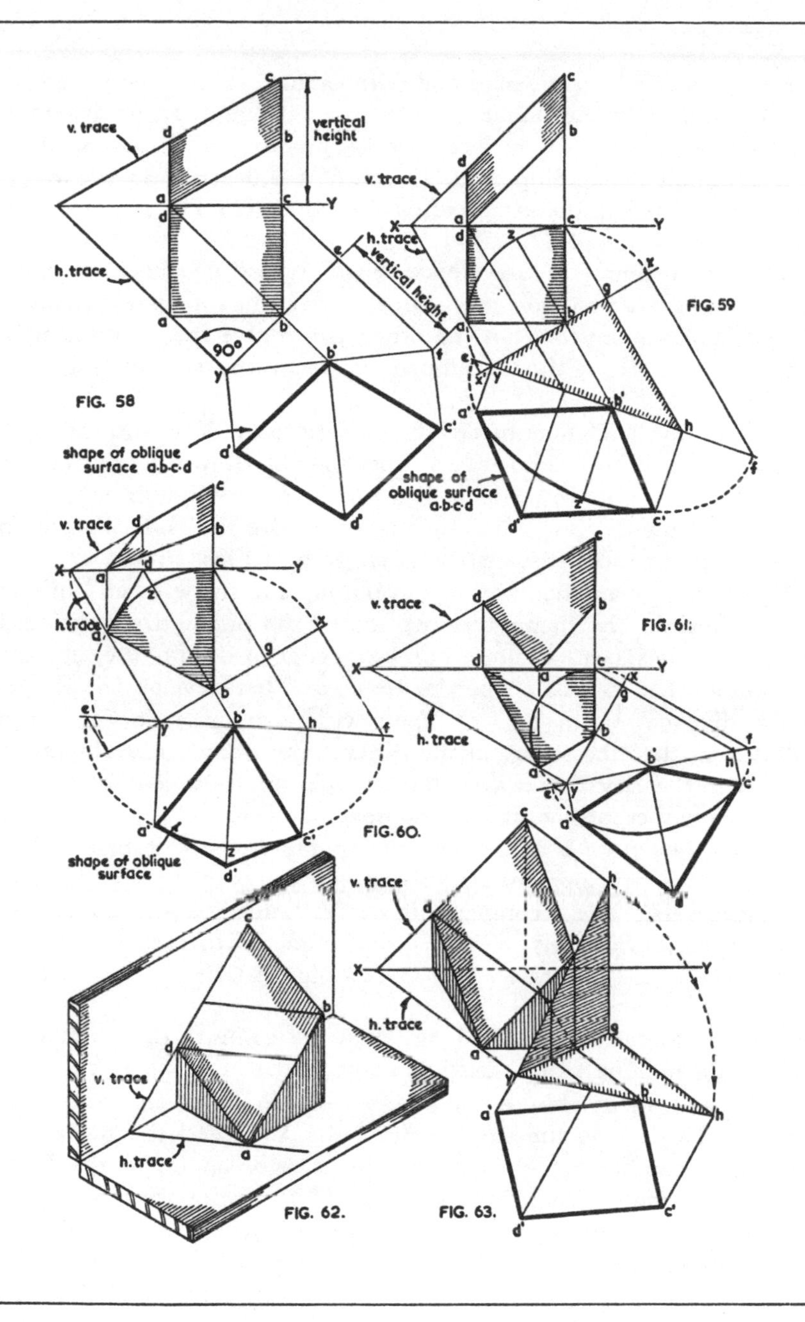

FIG. 58

FIG. 59

FIG. 60.

FIG. 61.

FIG. 62.

FIG. 63.

compass point in c in the plan and with radius c–a describe an arc to give a'' on the x–y line. Join a'' to c to give the length of the diagonal. To draw the shape of the surface, Fig. 54, first draw the line a–c. With compass point in a and open the length of a–b describe an arc in the direction of b. With the same radius and compass point in c make an arc in the direction of d.

Open the compass the developed length of a–d and b–c, and with the point in a make another arc to cut the arc at the top, giving point d. With compass point in c cut the other arc to give the position of b. a–b–c–d, Fig. 54, is the developed shape of the surface a–b–c–d, Fig. 53.

To develop the inclinations of the surface to the V.P. and the H.P. the vertical and horizontal traces must first be drawn. Continue the line c–d, Fig. 53, downwards until it meets the x–y line to obtain the V.T., and from T draw a line through a to give the H.T. To obtain the inclinations follow the instructions given in Figs. 50 and 51.

Fig. 55 shows another way of developing the shape of an oblique plane a–b–c–d. The shape is shown at the top of the drawings, and the surface has been swung round on edge a–d into the upright position so that its true shape can be seen. First develop the length of the diagonal, as in Fig. 53, then with compass point in a and radius a–c'' describe an arc in the direction of c in the development. From c in the elevation draw a line at right angles to a–d (the hinge line) to give the position of c in the development.

Draw a line from b in the elevation parallel to c–c to meet a line from c in the development parallel to a–d to complete the shape of the surface. As a–d is in contact with the V.P. this line is also the V.T. of the surface. Similarly, as a–b is in contact with the H.P. this is also the H.T. of the surface. Fig. 57 is a pictorial view of the surface.

Oblique planes for handrailing. Now we come to the method of developing oblique planes used in handrailing. Fig. 58 is the plan and elevation of an oblique plane a–b–c–d, similar to that shown in Fig. 56. To develop the surface draw the V.T. and the horizontal trace, also another line through b and at right angles to the H.T. Draw lines from b, c and d parallel to the H.T., and on the line from c, and, measuring from e, mark off the vertical height of the surface. This is taken from the elevation. Join f to y and from y, b' and f draw lines at right angles to y–f, making y–a' equal y–a, b'–d' equal

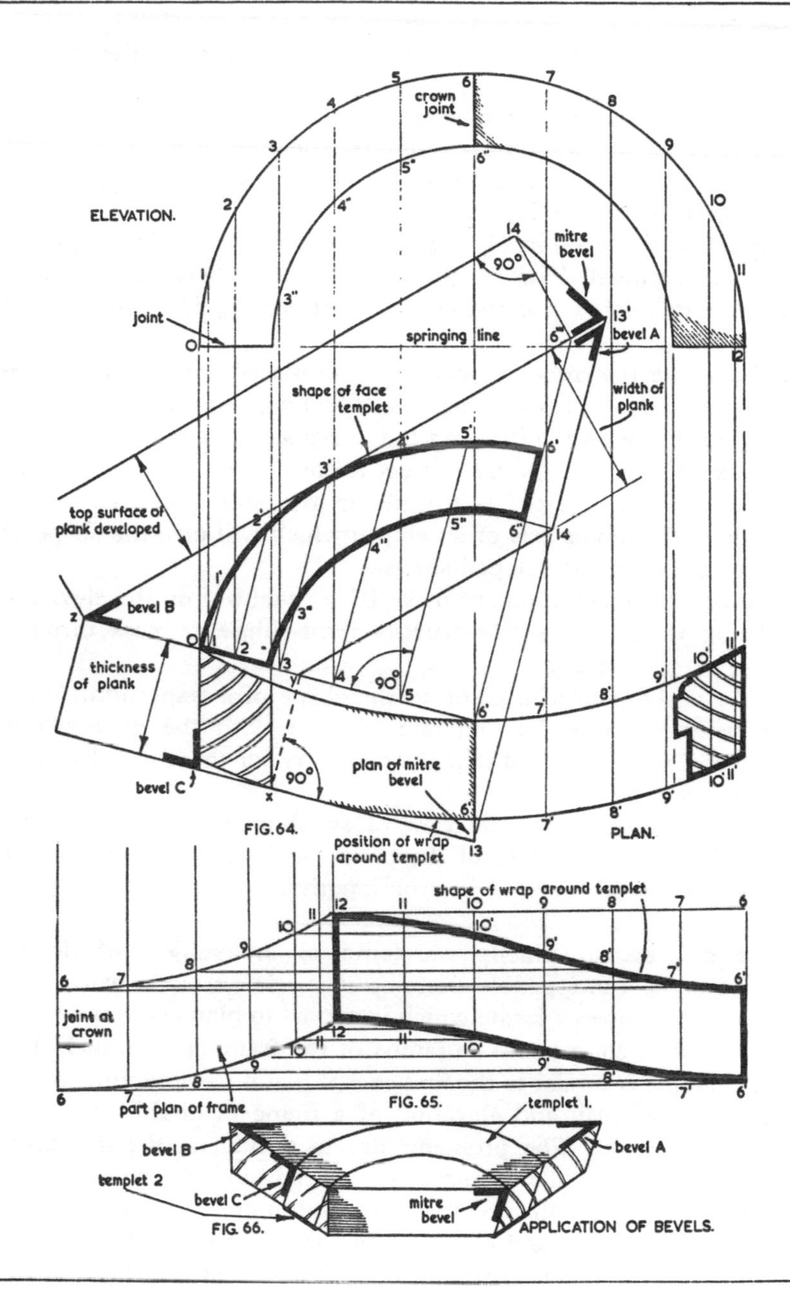

FIG. 64.

FIG. 65.

FIG. 66.

to b–d, and f–c′ equal to e–c. a′–b′–c′–d′ is the shape of the oblique surface.

Fig. 59 shows another and similar oblique plane developed, and when this is compared with the geometry shown in Fig. 58, it will be seen to be the same in almost every detail. Now study Fig. 63, where this geometry of the oblique plane is illustrated. In addition to the surface being developed in Fig. 59, the curve a–c has also been placed on the development. This is particularly useful because, in handrailing, the templets for applying to the timber are developed in this way.

It should be remembered that any line placed across the surface and parallel to the horizontal trace is a level line, and can be measured in the plan. Place on the drawing the curve a–c and extend it round to x and x′. From these points project it down to the y–h line to give e and f. The distance e–f is the major axis and b–z, seen in the elevation, is the minor axis of an ellipse which will give the shape of the curve a–c on the developed surface.

From b′ on the development make b′–z equal b–z on the elevation and, using these points, construct a semi-ellipse to pass through points e, a′, z, c′ and f.

Fig. 60 is the plan of a plane in the shape of a trapezium. If the reader compares this drawing and Fig. 61 with the drawings in Figs. 58 and 59 he will find that the geometry for these two drawings is exactly the same.

Fig. 62 is another pictorial view of an oblique plane, and shows that any line on the surface of the plane, so long as it is parallel to the H.T., is a level line and its true length can be seen in the plan.

Double curvature. Finally we come to the subject of double curvature or, as some people describe it, circle on circle work. The first example involves a frame which is curved in plan and has a semicircular head in elevation. The jambs of the frame are parallel. The second example is similar to the first except that it has radiating jambs.

Fig. 64 is the plan and elevation of a frame in double curvature with parallel jambs. The problems are to decide on the minimum thickness and width of the material required, and the shape of the templets necessary for shaping the head, and to develop any necessary bevels. Draw the plan and elevation to the dimensions of the frame and divide the elevation curve into any number of equal parts, say twelve. Draw a line to connect points O and 6′ in the plan, and another line parallel to this to just touch the outside edge of that half of the frame. The distance between these two lines represents the minimum thickness of the timber required to produce the two halves of the head of the frame.

Project vertical lines down from the points round one half of the elevation, to intersect with the o–6′ line in the plan. Draw lines from all these points at right angles to o–6′, and make the lengths of these lines equal to those in the elevation, measuring from the o–12 or springing line. Mark off, too, the heights of the 3″, 4″, 5″ and 6″ points on these lines. Freehand curves through these two sets of points will give the shape of the face templet required for the work.

The minimum width of the material is found as follows. Draw a line from x in the plan at 90° to the edges of the plank to give point y on the opposite side. Next draw a line from 13 in the plan parallel to 6′–6″ to meet a line parallel to o–6′ brought out from 6″, giving point 14. Also put in a line from y to 14 to give one edge of the plank. Draw another line parallel to y–14 to just touch the outside edge of the templet shape. The distance between these two lines is the minimum width of timber required for the head.

The mitre bevel is found by developing the shape of the top surface of the plank, and this is shown above the development of the templet.

When the setting out has been completed two pieces of timber will be required, one left and one right hand, shaped to y–14–13′–z, seen around the templet shape, each piece being equal in thickness as that seen in the plan. With the mitre bevel applied to the top end of each, the two halves can have the face templets applied to the front and back surfaces. Fig. 66 shows how these and the bevels are applied, and the two halves shaped to the outline of the templets. When this has been done the wrap-around templet can be applied round the outside edge of each half to obtain their shapes seen in the plan.

To develop the shape of the wrap-round templet, Fig. 65, project the points 6 to 12 around the edge of the elevation downwards to give points 6 and 6, 7 and 7, 8 and 8, etc. on the front and rear edges of the frame in plan. Project horizontal lines from all these points over towards the right and on the line from 12 step off six distances equal to those round half the elevation. Drop lines downwards from the points obtained to intersect with the horizontal lines brought over from the plan to give 11′ and 11′, 10′ and 10′, 9′ and 9′, etc. Freehand curves through these two sets of points will give the shape of the wrap-around templet.

The templet should be made from some thin, flexible material such as $\frac{1}{8}$ in. plywood. When placed correctly it should give the outline of the curve as seen in the left hand half of the plan, Fig. 64. Each half of the frame can then be shaped, and checking that the surface is always vertical when the piece being shaped is stood up on its lower joint surface.

Cuneoid problems. Fig. 67 is the plan and elevation of a semi-circular-headed frame, curved in plan and with radiating jambs. Only a brief outline of the geometry involved can be dealt with in this chapter. This frame head is based on the geometrical solid known as the cuneoid, see Fig. 70. The outline of the cuneoid in the plan, Fig. 67, is o–6–a–z. The elevation of the front face of the solid is the semicircle o–6–a, and its back edge is h–n.

To develop the bevels, templets, etc. draw the plan and elevation and divide one half of that drawing into, say, six equal parts. Draw a line across the o''–6'' points in the plan, and then another parallel to it to just touch the outside edge of the plan of the frame. The distance between these two lines is the minimum thickness of the timber required.

Drop lines from the points round the outside of the elevation down to the o–6 line (this is the front face of the cuneoid in the plan) to give points 1, 2, 3, etc. on this line. From z, the back edge of the cuneoid, draw lines through o, 1, 2, 3, etc. to give points o'', 1'', 2'', 3'', etc. on each edge of the plank in the plan. From these points draw lines at right angles to the plank edges, and make these lines equal to those in the elevation. The broken line in the elevation represents the inside edge of the frame and these various heights can also be placed on the appropriate lines in the templet development. Freehand curves through the points obtained will give the shapes of the two templets required for this type of frame. The outline of the plank which gives the minimum width, and the development of the mitre bevel, are made in the same way as for Fig. 64, but for accuracy the development of the plank width should be made on the convex side of the plan. To obtain point y draw a line from x at right angles to the plank edges to give y on the convex side of the frame.

Fig. 69a shows how the two templets are applied to each half of

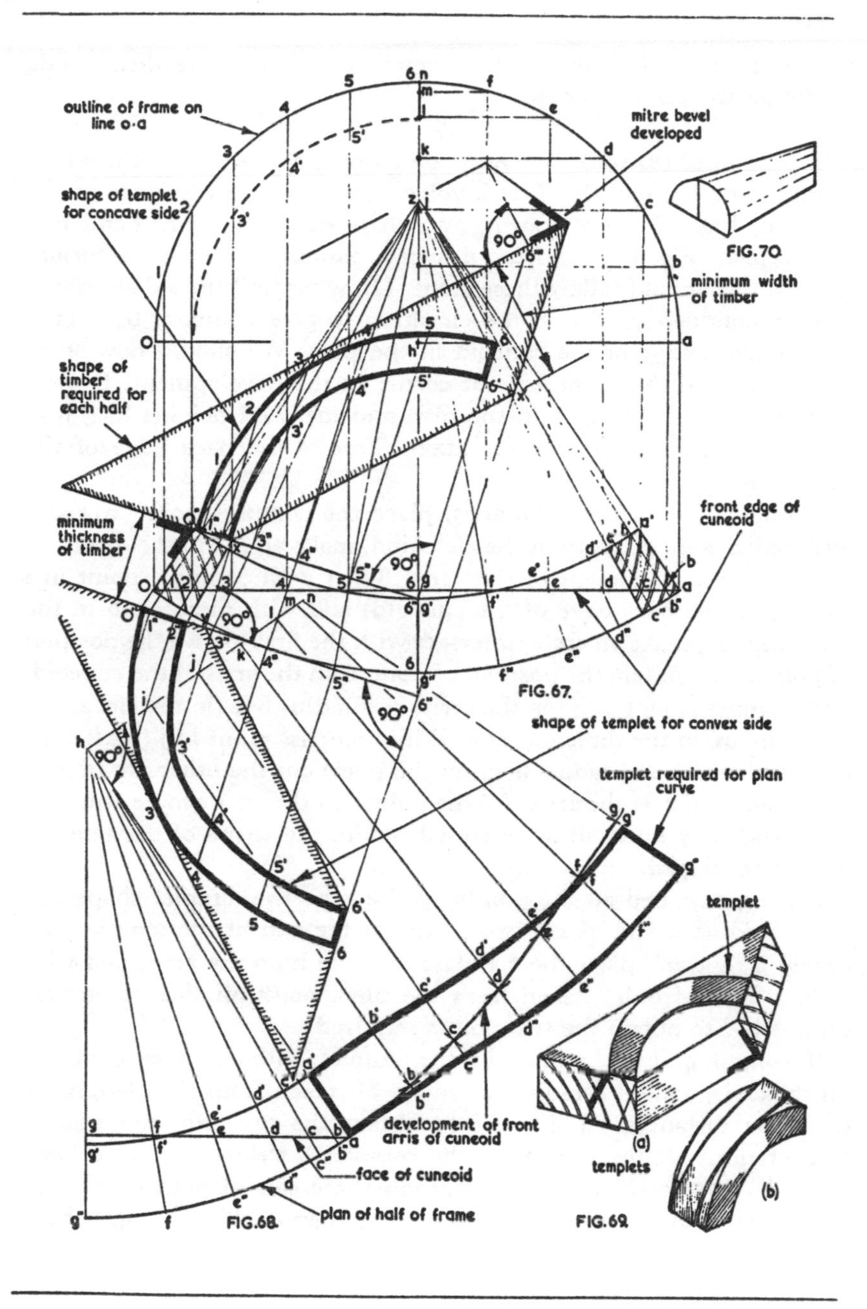
outline of frame on line o·a
mitre bevel developed
shape of templet for concave side
minimum width of timber
shape of timber required for each half
minimum thickness of timber
front edge of cuneoid
shape of templet for convex side
templet required for plan curve
templet
development of front arris of cuneoid
face of cuneoid
plan of half of frame
templets
(a)
(b)
FIG.67.
FIG.68.
FIG.69.
FIG.70.

the head, and at b is shown the wrap-around templet applied to one of the partly-shaped halves.

Wrap-around templet. We now come to the setting out of the wrap-around templet, and this first involves developing the surface of the cuneoid, Fig. 68. This drawing, which is really the right hand half of the plan, Fig. 67, has been drawn separately to avoid confusion. Briefly, one should follow these steps. Draw vertical lines downwards from points a–b–c, etc. on the elevation to give points a, b, c, etc., on the front edge of the cuneoid in the plan. We should now show the position of this front arris or corner in the development. Draw a line at 90° to a–h from h in the plan and mark off heights h, i, j, k, etc., equal to those in the elevation. This is the back edge of the cuneoid.

To obtain point b on the arris, place the compass point in a and with radius a–b, taken from the elevation, make an arc in the direction of b in the development of the arris. Then with compass point in i (this is on the back edge of the cuneoid) and with radius h–b in the plan, Fig. 68, make an arc to intersect with the first to give the position of point b. To obtain the position of point c on the arris of the cuneoid, with compass point in b (on the arris) and radius b–c (in the elevation) make an arc in the direction of c. With compass point in j (back edge of the cuneoid) and radius h–c (on the plan) cut the other arc to give the position of c on the arris. All the points on the arris can be obtained thus, and they must all be obtained before the shape of the templet can be developed.

Having obtained all the points on the arris the templet shape can be arrived at. a and a′ are two points at the end of the templet. To obtain b′ and b′′ place the compass point in b on the arris, and with radii b–b′ and b–b′′ taken from the plan, mark off these distances on line b–i to obtain the two points required.

To obtain points c′ and c′′ on the templet, with compass point in c on the arris, and with radii c–c′ and c–c′′ taken from the plan, mark off these distances on line c–j to obtain the two required points. Repeat until all the points on the templet development have been plotted. Two freehand curves through these sets of points will give the shape of the wrap-around templet. It is applied as in Fig. 69b.

20 Fixing Devices for Carpentry and Joinery

In recent years many devices have been introduced into the building industry so that items of carpentry and joinery can be securely fixed to all kinds of material in all kinds of situations. The *Rawlplug Company Ltd.* have done a great deal in this field, and the first part of this chapter has been devoted to many of their devices. Perhaps the most commonly used of their products are *Rawlplugs* (Fig. 3a) made from strong natural fibres which have been compressed to form a rigid tube. They are also treated to resist attacks by fungi and bacteria, and are extremely simple to use. If the hole is to be cut by hand a *Rawlplug* tool or *Rawldrill* similar to that shown in Fig. 1 is required; also a hammer. When the hole position is known, the tool is held in line with the required direction, with the point in the centre of the hole position. The top of the tool is tapped with the hammer, and as the tapping is continued, the tool is revolved until the required depth of hole is obtained. The hole should be deep enough to allow the whole of the plug to enter with the tip just below the surface. The screw can then be turned into the plug with a screwdriver.

For heavy concrete, granite, hard bricks and masonry, the *Rawlplug* tool is probably the best device to use for making a hole, but for comparatively soft materials such as light concrete, soft brickwork, stone and slate the holes are best made with a durium-tipped drill, Fig. 2, fitted to a portable electric drill, Fig. 7, Chapter 21. These tipped drills, of course, can also be used in a wheel-brace or hand-brace and produce good accurate holes. An important point is that the tool number must be the same as that of the plug. Thus when a No. 10 plug is being used, the tool must also be No. 10.

Another advantage of *Rawlplugs* is that screws other than wood screws can be used. Plugs as large as 4 in. long by 1 in. in diameter can be obtained, and it is evident that for that size of plug something like a $\frac{3}{4}$ in. coach screw would be required to justify so large a plug.

Where high temperatures occur near to the point where an item is to be fixed with plugs, or where moist conditions exist, it is wise to use a plug unaffected by such conditions. The plugs most suitable in such cases are white bronze metal plugs, Fig. 3b. They are used with wood screws in the same way as the normal plug, but will not carry the same loads as the fibrous plugs. The makers recommend that cadmium plates or stainless steel screws are used with white bronze plugs when full anti-corrosive protection is required. The large plugs can be used with coach screws. The largest white bronze plug available has a $\frac{5}{8}$ in. diameter and is 4 in. long.

Another special type of plug is the lead screw anchor, Fig. 3c, and should be used where ordinary metals are liable to be attacked by corrosive agents in the atmosphere. They are made from an alloy containing lead and antimony. The plugs are tapered internally to ensure expansion of the inner end of the plug, and they also have a flanged head to enable them to be used in hollow masonry. They are made up to size number 14, the largest being 2 in. long. It may be necessary to make an extra deep hole in masonry, and to accomplish this special 'long series' durium-tipped drills are available. These, Fig. 4c, give a drilling length of 6 in. and are made in sizes 12, 14, 16, 18 and 20. The extension type is another but similar drill which has a screwed shank to fit into an extension sleeve, Fig. 4a and b. The sleeves are 9 in. long and are screwed at both ends so that a number of them can be joined together to drill an extra deep hole. The extension-type durium drill is supplied up to 2 in. diameter; so also are extension sleeves to fit the drills.

Hollow wall fixings. It is sometimes necessary to fix items such as picture rails, skirtings, frames, etc., to hollow partitions or to some other form of hollow construction such as cladding. There are several devices which are extremely good for this work, and for vertical surfaces the gravity toggle, Fig. 5a, or the *Rawlanchor*, Fig. 8, can be used. The gravity toggle has a hinged bar on a swivel nut with one end heavier than the other so that it will always hang vertically. Thus when the toggle has been passed through a hole in a hollow wall the bar will drop into its vertical position and so prevent the device from being withdrawn, see Fig. 5c. In fixing, the screw must first be passed through the item to be screwed to the wall, and then entered into the

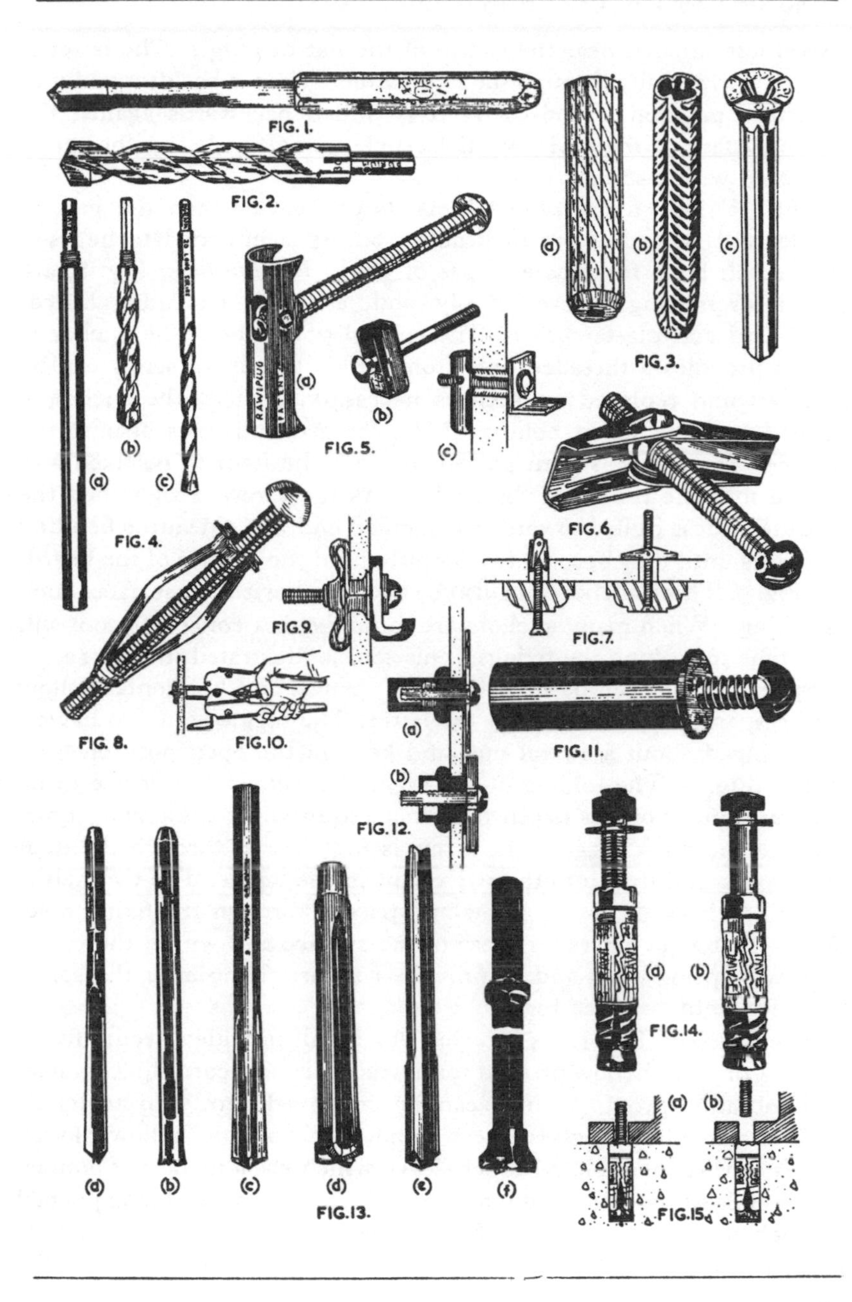
FIG. 1.
FIG. 2.
(a)
(b)
(c)
FIG. 3.
(a)
(b)
(c)
FIG. 4.
RAWLPLUG
(a)
(b)
(c)
FIG. 5.
FIG. 6.
FIG. 7.
FIG. 8.
FIG. 9.
FIG. 10.
FIG. 11.
(a)
(b)
FIG. 12.
(a)
(b)
(c)
(d)
(e)
(f)
FIG. 13.
(a)
(b)
FIG. 14.
(a)
(b)
FIG. 15.

swivel nut situated near the centre of the bar or toggle. The latter is pushed horizontally through the fixing hole. When it has dropped into a vertical position in the cavity it is pulled backwards against the inside surface of the cavity by the article being fixed, and the screw tightened with a screwdriver.

The 'H' type toggle, Fig. 5b, is much heavier than the gravity toggle, and is made from malleable iron. It is intended to be used with coach bolts for heavier types of work. *Rawlanchors*, Fig. 8, are useful for making fixtures in plywood, asbestos, insulating board, hardboard and plaster board. The special point about the anchor is that it provides a threaded hole from which the fixing screw can be removed and replaced as often as necessary. In use, the anchor is pushed into the fixing hole until the flanged end rests against the surface. The screw is then passed through the item to be fixed and turned into the thread of the anchor. As the screw is tightened the threaded nut is pulled towards the flanged end, the side arms bending outwards until they become two loops behind the surface of the board, see Fig. 9. These anchors can also be used for horizontal surfaces such as ceilings. When many anchors are to be fixed, a collapsing tool will save time in making the fixings. This tool is illustrated in Fig. 10.

Spring toggles can be used for both vertical and horizontal hollow surfaces, and they are ideal for the latter. The toggle is in two halves, each hinged about a swivel nut, and kept in the open position by a spring, Fig. 6. The folding of the toggle bar allows the device to be used in cavities of less depth than that required for the gravity type. As with the other toggles, the screw is first passed through the item to be fixed, and then into the swivel nut in the toggle. The two halves are folded back on to the screw and passed through the fixing hole. When the toggle halves are clear of the surface and within the cavity they will spring open and so provide a secure fixing after the screw has been tightened, see Fig. 7.

The *Rawlnut*, Fig. 11, is a device which will provide a secure fixing in thick or thin, hollow or solid materials. It is non-corrodible, water- and vibration-proof. Fixings can be made efficiently in materials such as thin plastics, glass, sheet metal, plywood and hollow blocks, etc. It consists of a tough rubber sleeve which has a metal nut bonded in one end. At the other end is a flange which prevents its being pushed through the fixing hole. The *Rawlnut* is pushed into the fixing hole up to its flange and the article to be fixed placed in position over the

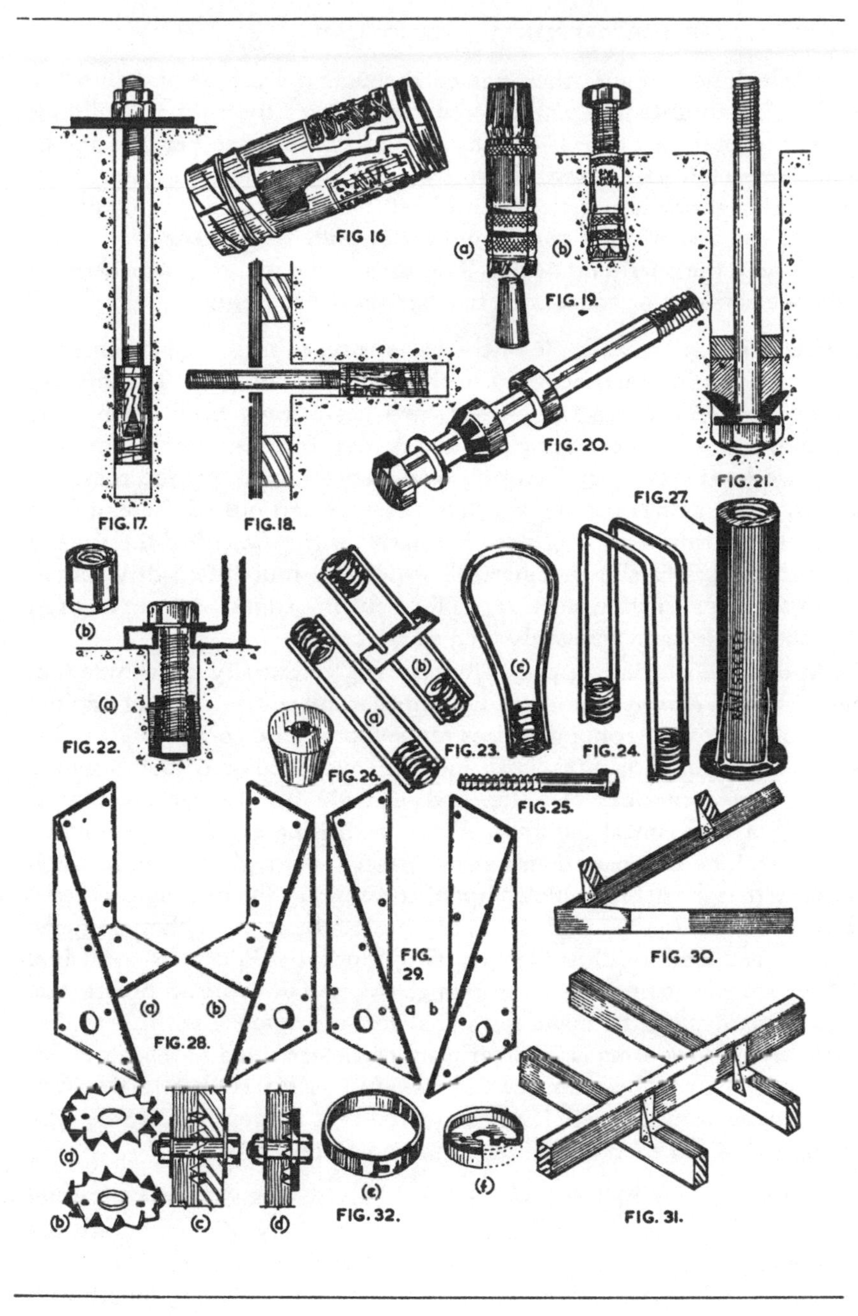
FIG 16
(a)
(b)
FIG. 19.
FIG. 20.
FIG. 21.
FIG. 17.
FIG. 18.
FIG. 27.
(b)
(a)
FIG. 22.
(b)
(a)
(c)
FIG. 23.
FIG. 24.
FIG. 26.
FIG. 25.
FIG. 30.
FIG. 29.
a b
(a)
(b)
FIG. 28.
(a)
(b)
(c)
(d)
(e)
(f)
FIG. 32.
FIG. 31.

Rawlnut. The screw is then passed through the article and into the device. The tightening of the screw compresses the rubber, building it up into an inner collar larger than the fixing hole, see Fig. 12a and b. The completed fixing is vibration free and is airtight and waterproof. The fixing screw is electrically insulated from any surrounding metal.

Fig. 12a and b show two kinds of *Rawlnut*. At a is the sealed-end type, and b the open-end device. The advantages of the first are obvious where electricity or some corrosive agents are present.

Drills. Fig. 13 shows some of the various types of drills available for cutting holes in materials used in the building industry. The first, a, is the *Rawldrill,* a description of which has already been given. The second, b, is a tubular boring tool and is ideal for boring holes through brickwork. It has several cutting teeth which break up the masonry quickly, the debris passing through the tool and out of the exit hole near the cutting end. It is used similarly to the *Rawldrill* (a) already described. At c is shown a stardrill which is a four-fluted drill forged from octagonal section steel. It has four cutting edges, and can be used for cutting holes in materials such as concrete.

At d is the durium-tipped hammer drill, essentially a machine tool suitable for *Kango* Hammers and similar machines. It will rapidly cut holes in tough concrete. Sizes range from $\frac{7}{16}$ in. to $1\frac{3}{4}$ in.

The cruciform drill (e), is an inexpensive percussion tool designed for a limited number of holes, and is made from rolled steel. It is used in a mechanical hammer. At f is a drilling bit for a pneumatic machine. The tool has an air passage passing through its length which ensures that the debris is blown out of the holes as the drilling proceeds. Fast drilling into the toughest of concrete can be made with these tools.

Fig. 14a and b illustrate the two types of *Rawlbolts* available. These are excellent devices for fixing bolts to brickwork, concrete, etc. Secure fixings can be made in the hardest of masonry with them. No cementing or grouting is necessary, only a clean drilled hole is required. The hole having been made, the *Rawlbolt* is inserted, and the item being fixed is positioned. The nut or the bolt (depending on the type being used) can then be tightened to complete the fixture.

Rawlplugs can be used for fixing items to floors, walls or other surfaces, and heavy equipment such as machinery, pipework, etc. are efficiently secured to these surfaces with *Rawlbolts,* Fig. 14. The shield or outer shell is made from malleable iron and is divided lengthways into four segments. These are interlocked with each other but free to expand outwards under pressure from a tapered expander, which is drawn into the inner end of the shell by tightening the fixing bolt. They are made in sizes from $\frac{3}{16}$ in. to 1 in. in diameter, and are capable of supporting a weight of a few pounds up to several tons.

The loose-bolt type, Fig. 14b, enables the fixing bolt to be removed at any time. The shell contains the expander nut. Its purpose is to provide a fixing into which the bolt can be entered after the parts to be secured are in position. The fixing bolt is passed through the item to be secured and into the *Rawlbolt,* and is tightened with a spanner.

With the bolt-projecting type, Fig. 14a, the fixing bolt is the expander and cannot be removed. As the bolt projects from the masonry, this type is ideal for wall fixings. Fig. 15 shows the two types being used for the same job. In addition to the two types of *Rawlbolts* described above, accessories to these bolts can be obtained in the shape of hook and eye bolts, pipe hangers, pipe clips, etc.

Where it is necessary to make a fixing to carry heavy loads in a material that is weak near its surface but which has a hard core, *Rawlplug Duplex* anchors (Fig. 16) are the answer. These are three-segment anchors, and give parallel expansion over the whole of the length. With this type of expansion the stress applied to the stud is transmitted over a much greater area of the hole surface, resulting in a strong grip in a material of indifferent quality. The stud or fixing bolt is first screwed into the anchor until it just starts to expand and then, by means of the stud, is pushed into the hole to the required depth. Rotation of the stud will expand and fix the anchor, which is then ready to carry its full load.

Figs. 17 and 18 show two of these anchors in use. Fig. 17 could be details around a machine bed, and Fig. 18 where some heavy item has to be fixed to a wall which has panelling fixed in position. The makers recommend that the studs to *Duplex* anchors should be tightened to a given torque. The danger of overstressing the studs is eliminated and also gives an assurance that the studs are tight enough, and that the fixing has established a good grip in the masonry. The torque for these studs can range from 60 in./lb. for a $\frac{1}{4}$ in. diameter stud up to 250 ft./lb. for a 1 in. diameter stud.

For a fixing in good quality brickwork, medium density concrete, stonework and granolithic concrete, *Drilanchors* could be used, Fig. 19a. In order to drill the hole, the body is first used as a drill, either as a hand tool or by being placed in the chuck of a mechanical hammer. The anchor is then withdrawn and the expander inserted into the cutting end, Fig. 19a. The assembly is then re-entered into the hole and the body is driven down over the expander until the chuck or hand tool is flush with the surface of the masonry. This expands the lower end of the anchor, providing a firm grip. A sharp sideways jerk of the driving tool will snap off the head of the device, as in Fig. 19b, ready to take the fixing bolt.

Where bolt fixing is required in wet conditions the *Rawlplug caulking bolt anchor* is recommended, Fig. 20. This is fixed by the action of caulking, and is made with bolts from ¼ in. to 1¼ in. diameter. It consists of a hard-chilled iron or steel segmented cone cast integral with a special lead alloy crown. In larger sizes the crown is made in two parts to ensure correct caulking. Each is caulked separately. The action of caulking compacts the crown, driving the edges of the steel segments into the masonry, and so obtaining the initial grip, Fig. 21. As the bolt is tightened the segments are driven farther into the masonry, thus increasing the hold. The compacting of the crown effectively seals the fixing cone against the ingress of water or corrosive agents. Caulking tools are supplied for all sizes of bolt anchors.

Rawltamps, Fig. 22b, are all-metal, non-ferrous fixings designed to give rapid and secure fixing for bolts in masonry. They are installed by caulking in a similar manner to bolt anchors but are intended for shallow holes. Because of this they are ideal in conditions where deep boring is to be avoided. The *Rawltamp* has a metal nut of special conical shape, flanged at one end, and having cast around it a cylindrical shield of lead alloy to permit caulking, see Fig. 22b. Caulking tools are available for these devices.

Rawlsockets, Fig. 27, are used in concrete constructions where it is possible to determine the positions of the holes before the concrete is poured. Made of malleable iron they provide a fixing for bolts from ¼ in. diameter up to 1¼ in. diameter. Normally *Rawlsockets* are secured to the framework by a bolt passed through from outside, but, where necessary, nailing plates can be fixed to the inside surfaces of the formwork and the sockets fixed to the plates. The nailing plates and holding screws are recoverable when the formwork is dismantled.

Other devices for concrete work are *Rawlties,* Fig. 23a and b, *Rawloops,* 23c, and *Rawlhangers,* Fig. 24. The use of these is explained in Chapter 5.

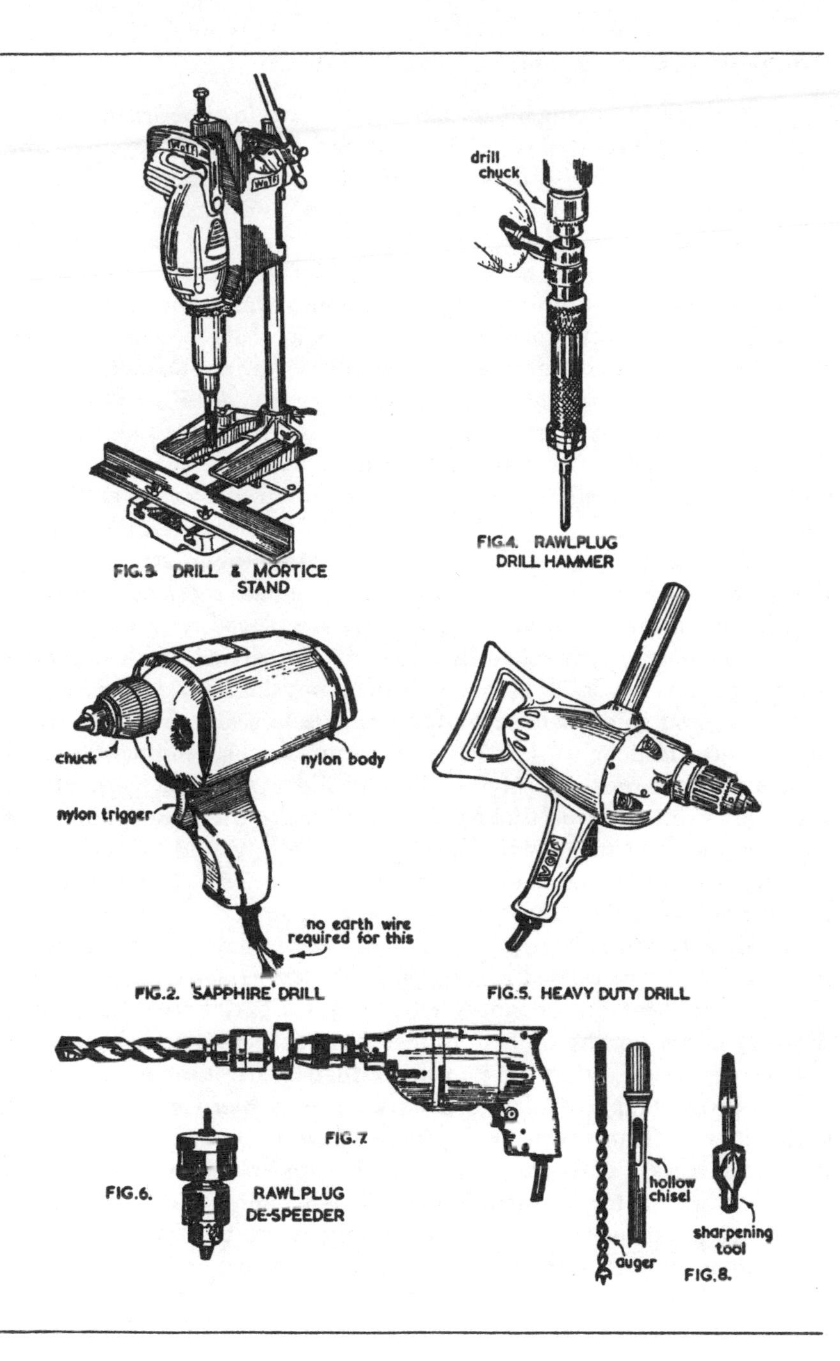

FIG. 3. DRILL & MORTICE STAND

FIG. 4. RAWLPLUG DRILL HAMMER

FIG. 2. 'SAPPHIRE' DRILL

FIG. 5. HEAVY DUTY DRILL

FIG. 7

FIG. 6. RAWLPLUG DE-SPEEDER

FIG. 8.

Another company which has done a great deal in experimental work in the building industry is *McAndrews and Forbes Ltd.* Amongst the devices they have had a great deal to do with are the Trip-L-Grip anchors, Fig. 28a and b and Fig. 29a and b, and timber connectors, Fig. 32a, b, c, d, e and f. The anchors, Figs. 28 and 29, have done a lot in recent years to cut out the uneconomical joints one has to prepare in such framed work as trimming around fireplaces and stairwells, and for such jobs as connecting purlins and binders in roofs to other timbers, Figs. 30 and 31. These drawings show only a few applications.

Timber connectors are made in three types. At Fig. 32a is the double-sided toothed variety, and at b the single-toothed kind.

A split-ring connector is given at e and at f the shear-plate connector. Connectors largely eliminate complicated joints in roofing and other similar work. Instead of the compression joints used years ago when large sizes of timbers were common, today timber structures are designed to be made from comparatively small sections, many of which have to be in tension. When bolts are placed near the ends of timbers in tension and when heavy loads are carried there is always a danger that the bolt under load will shear out the timber immediately behind it, and so render the bolted joint useless. To avoid this connectors are used between each pair of timbers, spreading the load over a much larger area of the timber, Fig. 32c. The teeth of the connectors bite into the timber fibres and make the joints capable of carrying much greater loads than they normally would be capable of bearing. Where metal and timber are to be bolted together the single-sided connector must be used, Fig. 32d.

Toothed connectors are used for reasonably small jobs. Where a heavy load is to be carried it is necessary to use split-ring connectors, Fig. 32e, or shear-plate connectors, Fig. 32f. Special tools have to be used for preparing the timbers to take both split-ring and shear-plate connectors because they have to be bedded in the timbers correctly. The split-ring type requires a groove in each piece of timber with which it is in contact, and these grooves have to be such that the ring has to be forced outwards slightly to be able to enter the grooves. This will make for a much more efficient joint. The shear-plate connector is used where timber and metal have to be secured together in a similar manner to the single-toothed connector.

21 Portable Electric Tools

In the future carpenters and joiners will be expected to make more use of machinery than they do at present. The cutting of mortices and tenons by hand, and many other tedious operations are fast becoming old-fashioned, and in recent years the manufacturers of portable woodworking tools have made great strides in developing these machines. Naturally, portable tools have their limitations compared with traditional woodworking machines, but they have one great advantage in that they can be taken to the job. Bench joiners are able to take the drudgery out of bench work with the use of electric planers and routers, etc., and the carpenter on the site can use tools such as the portable electric saws for cutting roof timbers, floor joists, etc. Portable tools thus speed up the work and so help in keeping prices down.

Portable saws. Fig. 1 shows a *Wolf* electric circular saw. This has a cutting capacity of approximately 2½ in., but larger models are available. These tools can be used for cross-cutting and ripping timber. The guard is spring-loaded, and when the tool is removed from the timber it springs back to cover the saw teeth. The table can be tilted over to enable the saw to make bevelled or oblique cuts. A fence is also provided for ripping, and a pointer at the front enables the operator to keep to a line marked on the material.

Various types of blades are obtainable, including blades for cutting wallboard, hardboard, aluminium and corrugated sheets, and abrasive discs are available for cutting asbestos sheets, stone, brick, etc.

Drill. Probably the most-used portable tool is the drill. Fig. 2 shows a view of the *Wolf Sapphire,* which is the first British fully-insulated drill. As it is 100 per cent fully insulated there is no need for earthing this particular model. It has an all-nylon body and its performance is excellent. It is available in three sizes, $\frac{1}{4}$ in., $\frac{5}{16}$ in. and $\frac{3}{8}$ in. A stand can be obtained similar to that seen in Fig. 3 to cradle the drill when used for vertical drilling. The stand can also be used for cutting mortices in timber, special hollow chisels being obtained for this purpose.

Fig. 8 shows a hollow chisel used for morticing. The auger which passes through the centre of the chisel cuts the timber and passes the chips up and out through the opening in the side of the chisel. The auger is sharpened with a file in a similar manner to a joiner's twist bit, and the chisel is kept in condition with the tool shown to the right.

There are several items which can be used with electric drills, the drill hammer, Fig. 4, being one of these. This is an appliance which can be used when drilling fixing holes in masonry. It fits into the chuck of the drill, and a hammer effect is produced which is similar to that described in Chapter 20.

When a masonry drill is used for drilling holes in brickwork, etc., a de-speeder can be fitted for cutting down the r.p.m. of the tool, see Figs. 6 and 7. Fig. 5 is another drill which is capable of heavy-duty work, and of which common sizes can go up to a $\frac{1}{2}$ in. capacity.

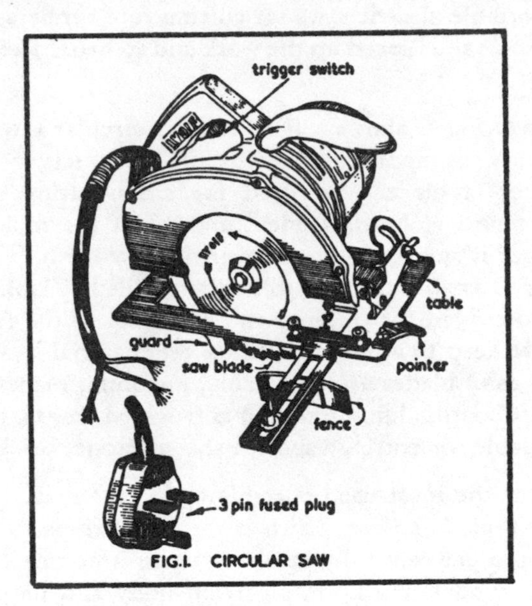

FIG. I. CIRCULAR SAW

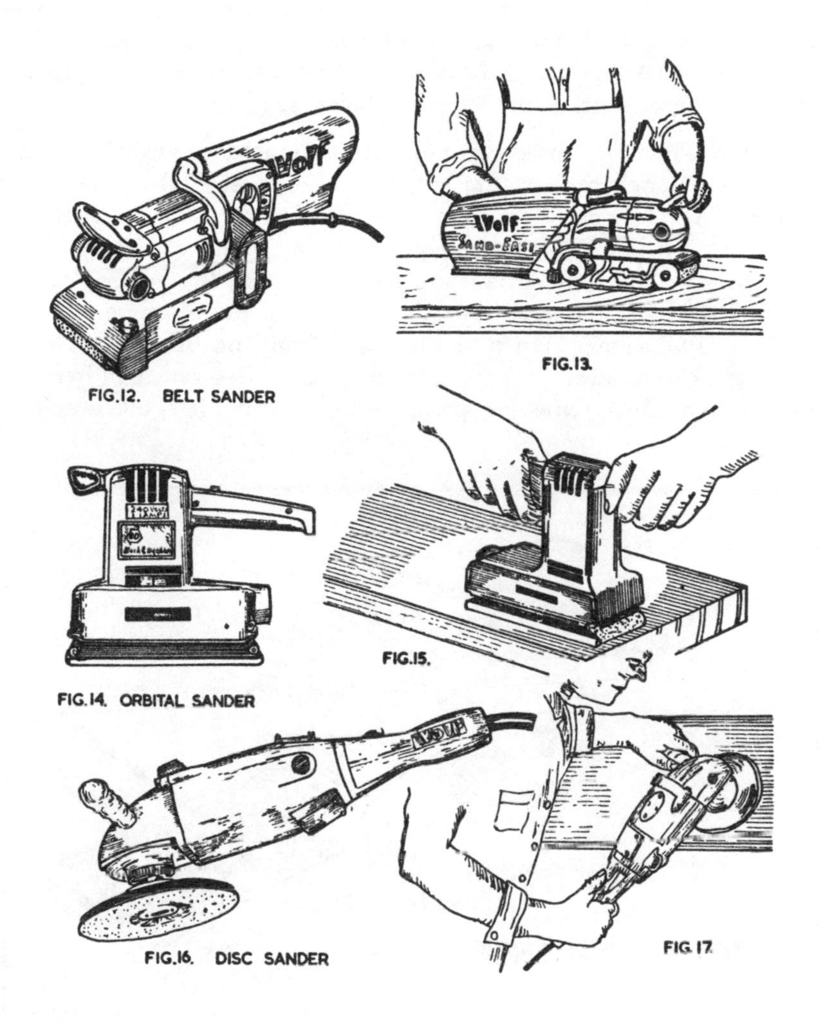

FIG. 12. BELT SANDER

FIG. 13.

FIG. 14. ORBITAL SANDER

FIG. 15.

FIG. 16. DISC SANDER

FIG. 17.

Electric plane. A useful tool, especially for joiners at the bench, is the electric planing machine, Fig. 9. This is an illustration of a *Tarplaner,* and can be used for work for which the jack plane is best suited. The machine has 4 in. wide cutters and it has a maximum depth of cut of $\frac{1}{8}$ in. It is easily and quickly adjusted. By moving the planer over the work slowly a fine finish can be obtained. The sharpening of the cutters should be done with a special tool, Fig. 11a. This is called a lapping tool, and enables the blades to be resharpened at the correct angle. A fence is also available which can be used for planing narrow surfaces such as the edges of doors, see Fig. 10.

Sanders. There are three types of sanding machines available to the woodworker; belt sander, Fig. 12, orbital sander, Fig. 14, and disc sander, Fig. 16. The belt sander is best suited for long flat surfaces. It can produce an accurate and good finish. It has a built-in dust-extracting system which transfers the dust from the surface of the wood to the cloth bag seen towards the back of the machine.

The orbital sander shown in Fig. 14 is from the *Black and Decker* range and can be used for producing a good surface on small items of joinery as well as removing paint and preparing irregular surfaces

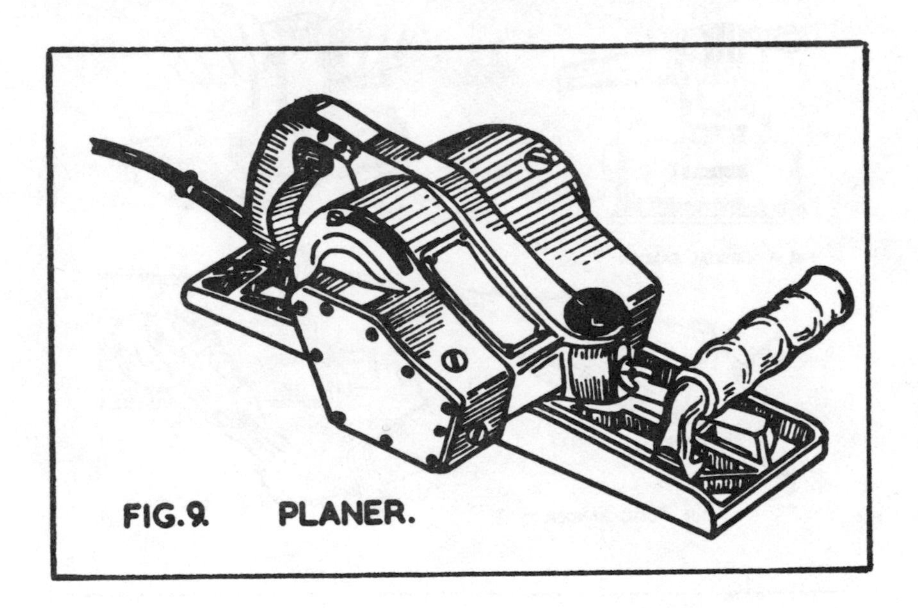

FIG. 9. PLANER.

such as floors, etc. The tool has an orbital motion, the plate holding the abrasive paper revolving very quickly, as fast as ·4,250 r.p.m. The diameter of the circle is around $\frac{3}{16}$ in. Abrasive papers can be changed quickly and easily.

The disc sander is for fast cutting. The abrasive disc fits over a rubber sanding head which allows uneven surfaces to be cleaned up. It is particularly useful in cleaning off synthetic resin adhesive on glulam work, prior to cleaning up the timber component with an electric planer. The belt and disc sanders are from the *Wolf* range.

Jigsaw. The *Black and Decker* jigsaw, Fig. 18, has a reciprocating action, and can cut timber up to a thickness of $1\frac{1}{2}$ in. It is useful for cutting out plywood and solid wood shapes, and is capable of doing almost any kind of work a bandsaw is able to do.

Stanley-Bridges produce some good portable tools, and the all-purpose saw, Fig. 20, is one of these. Blades can be obtained for cutting various metals as well as timber. Two examples of *Stanley-Bridges* portable routers are seen in Figs. 21 and 22. The first, Fig. 21, the *Stanley 'Spotlight'* $\frac{7}{8}$ h.p. router, has a motor which gives 23,000 r.p.m. Another router in the same range has a $1\frac{1}{4}$ h.p. motor. These machines give an accuracy to within ·004 in. The cantilever base leaves the work area open and gives excellent visibility. The on-off switch, when in the 'off' position, automatically locks the shaft and it is simple to change the cutters. *'Spotlight'* routers have a spotlight which illuminates the working area.

The second of the *Stanley* routers, Fig. 22, is the $\frac{1}{4}$ h.p. lightweight

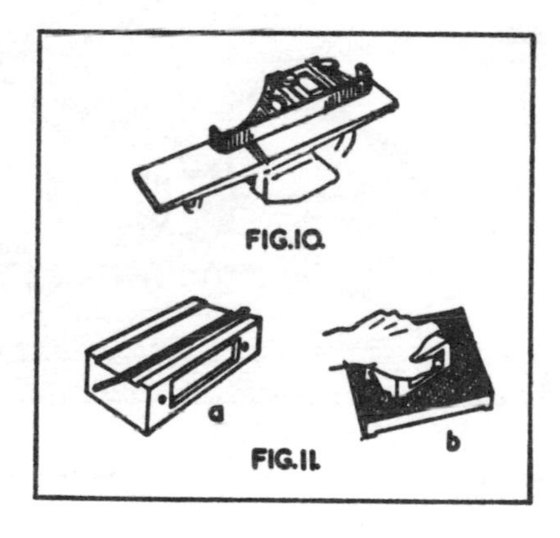

FIG.10
FIG.11

model. It has a rev. speed of 27,000 and produces high quality work. It weighs only 3¼ lb., and has a safety shaft-locking bottom. It also has a threaded motor housing and base giving easy and accurate adjustment of depth of cut. Router guides can be obtained which enable straight or circular cuts to be made. Other accessories available are groovamatic guides which extend the cutting widths of grooving bits, trammel points, and rods for cutting circles, and sub bases for use for work with templets.

Fig. 23 is an illustration of a *Centec* router, and Fig. 24 shows a few of the cutters which are available for this machine. As the cutters show, routers can do many jobs, including grooving, bevelling, chamfering, moulding and dovetailing.

The *Centec* router has a 1⅛ h.p. motor and a rev. speed of 18,000. It has many accessories available such as a straight guide which allows grooving to be done up to 9½ in. from the edge of the timber. A guide for circular faces is also obtainable. There is also a cross-cutting

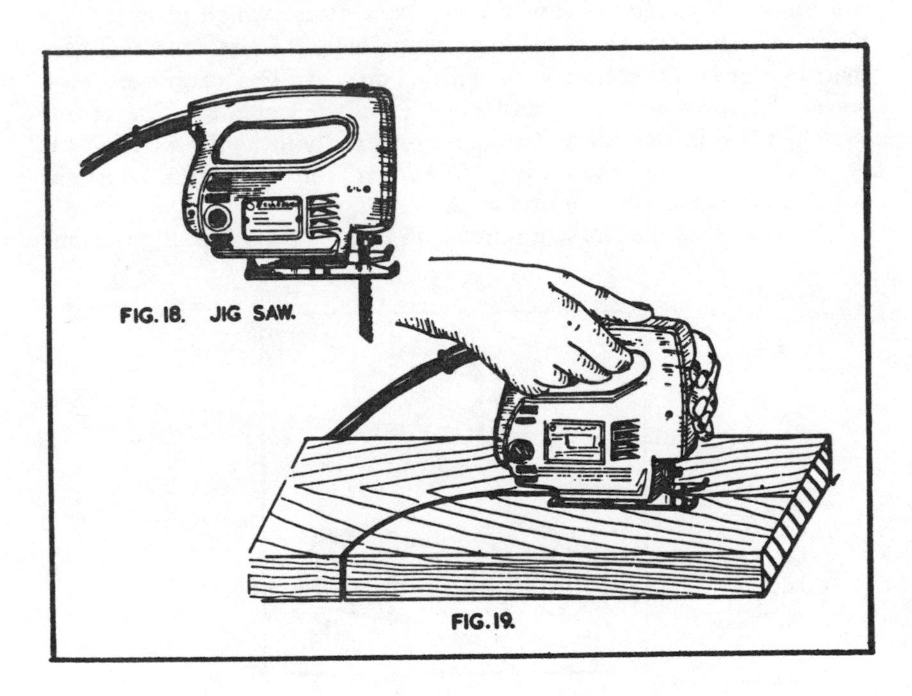

FIG. 18. JIG SAW.

FIG. 19.

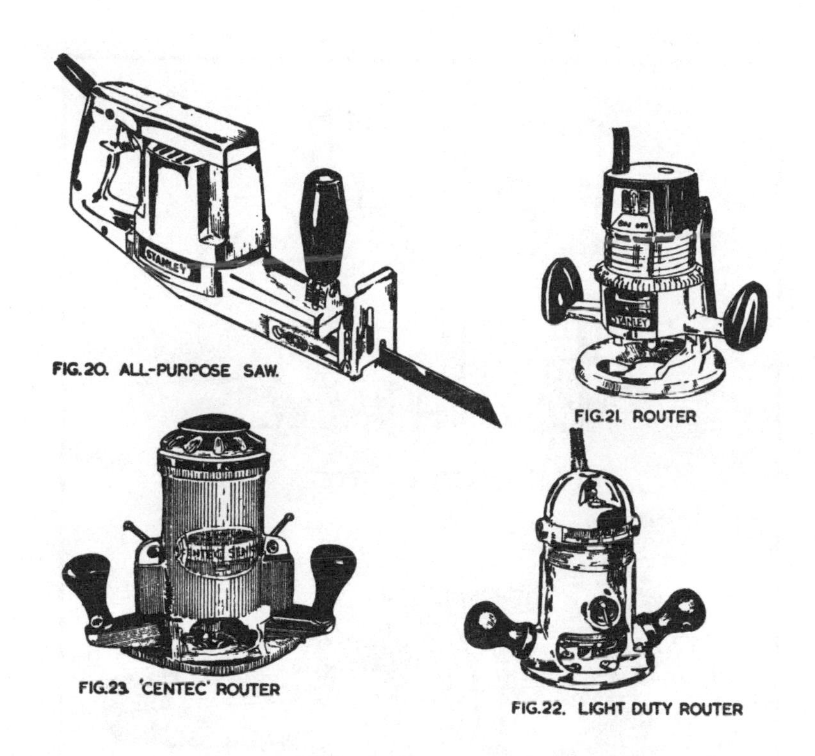

FIG. 20. ALL-PURPOSE SAW.

FIG. 21. ROUTER

FIG. 23. 'CENTEC' ROUTER

FIG. 22. LIGHT DUTY ROUTER

attachment which allows grooves to be cut accurately across a board up to 24 in. wide. There are also staircase recessing and dovetailing attachments, and a stand to which the router can be fixed so that it

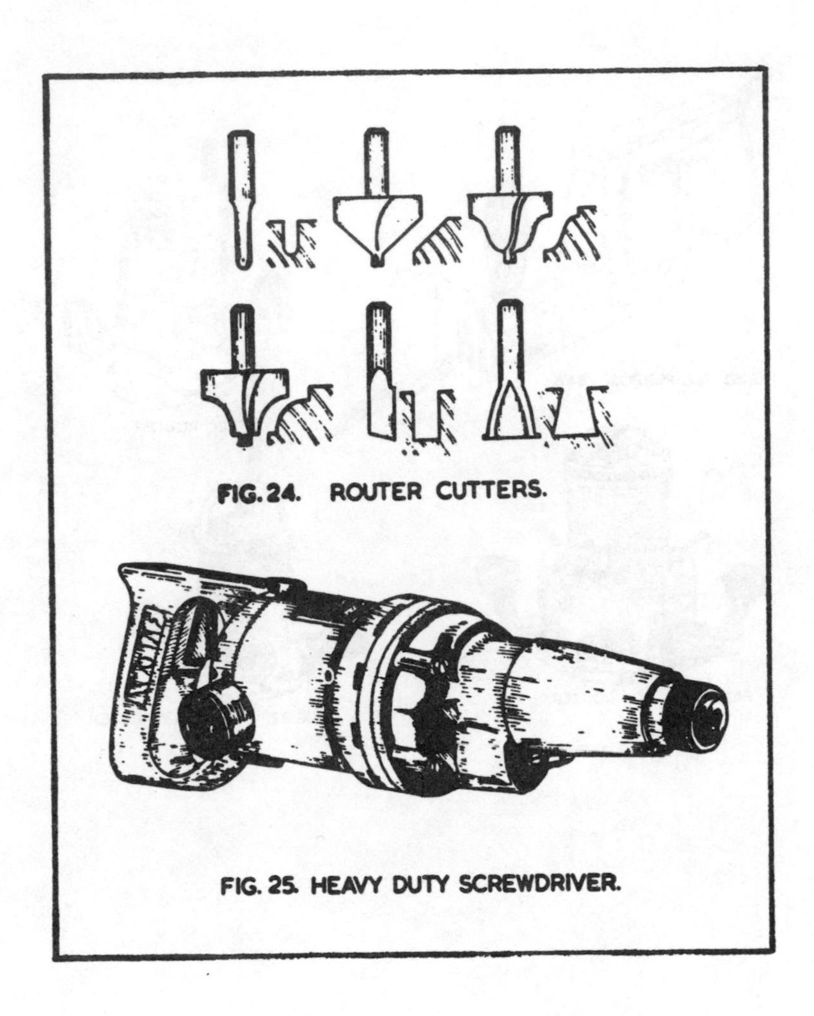

FIG. 24. ROUTER CUTTERS.

FIG. 25. HEAVY DUTY SCREWDRIVER.

can be used as an over-arm machine with an 18 in. by 18 in. table on which the work rests.

Depth of cut is set by unclamping the two thumb levers and turning the large dome nut on top of the machine body. One complete turn moves the cutter $\frac{1}{16}$ in. When the cutter has been adjusted to the correct position the thumb clamps are re-tightened. Cutters are supplied with $\frac{1}{4}$ in. and $\frac{3}{8}$ in. shanks.

Electric screwdrivers. This *Wolf* heavy-duty tool, Fig. 25, is ideal when repetition screwdriving is involved. This machine will drive a screw accurately to a pre-set tension at high speed. A sensitive clutch, adjustable to suit the screws being used, ensures against over-running, burring of screw heads and damage to surfaces. The screwdriver bit, not shown in the drawing, automatically finds the screw slot and idles until the pressure is applied. A reversing switch enables the screws to be extracted.

The user of electric appliances must realize that safety is an important factor, as carelessness can lead to serious accidents. A healthy respect for electricity must be maintained at all times. For guidance to purchasers and users of electric tools the following hints are given:

1. Make sure that the voltage indicated on the machine is correct for the power supply available.
2. The tool must be connected to the power point in the correct way. This should be done by a skilled and qualified person. Extension leads must be connected to the existing leads of the machine by proper plugs and sockets. The correct size cable must always be used.
3. No machine should be used for work which it is not capable of doing efficiently.
4. Tools should always be disconnected from the supply before adjustments are made.
5. Never carry or drag a machine along by its cable. This can either damage the cable or interfere with the connections.
6. Be sure that cutters, blades, etc. are kept sharp, as blunt cutters, etc. will cause the machine to stall. If a machine does stall it should be switched off and restarted under a lighter load.
7. Wear goggles when using grinders, sanders, etc. Protective clothing should also be worn at all times.
8. Never put a machine down on to a bench while it is running. Wait until it stops.

Index